Addition and Subtraction

Money, Miles, and Large Numbers

Grade 4

Also appropriate for Grade 5

Karen Economopoulos

Jan Mokros

Joan Akers

Susan Jo Russell

Developed at TERC, Cambridge, Massachusetts

Dale Seymour Publications®
Menlo Park, California

The *Investigations* curriculum was developed at TERC (formerly Technical Education Research Centers) in collaboration with Kent State University and the State University of New York at Buffalo. The work was supported in part by National Science Foundation Grant No. ESI-9050210. TERC is a nonprofit company working to improve mathematics and science education. TERC is located at 2067 Massachusetts Avenue, Cambridge, MA 02140.

This project was supported, in part,
by the
National Science Foundation
Opinions expressed are those of the authors
and not necessarily those of the Foundation

Managing Editor: Catherine Anderson
Series Editor: Beverly Cory
Revision Team: Laura Marshall Alavosus, Ellen Harding, Patty Green Holubar, Suzanne Knott, Beverly Hersh Lozoff
ESL Consultant: Nancy Sokol Green
Production/Manufacturing Director: Janet Yearian
Production/Manufacturing Coordinator: Joe Conte
Design Manager: Jeff Kelly
Design: Don Taka
Illustrations: Margaret Sanfilippo, Rebecca Krug
Cover: Bay Graphics
Composition: Archetype Book Composition

This book is published by Dale Seymour Publications®, an imprint of Addison Wesley Longman, Inc.

Dale Seymour Publications
2725 Sand Hill Road
Menlo Park, CA 94025
Customer Service: 800-872-1100

Order number DS43896
ISBN 1-57232-749-9
1 2 3 4 5 6 7 8 9 10-ML-01 00 99 98 97

Printed on Recycled Paper

T E R C

INVESTIGATIONS IN NUMBER, DATA, AND SPACE®

Principal Investigator   Susan Jo Russell

Co-Principal Investigator   Cornelia C. Tierney

Director of Research and Evaluation   Jan Mokros

Curriculum Development

Joan Akers
Michael T. Battista
Mary Berle-Carman
Douglas H. Clements
Karen Economopoulos
Ricardo Nemirovsky
Andee Rubin
Susan Jo Russell
Cornelia C. Tierney
Amy Shulman Weinberg

Evaluation and Assessment

Mary Berle-Carman
Abouali Farmanfarmaian
Jan Mokros
Mark Ogonowski
Amy Shulman Weinberg
Tracey Wright
Lisa Yaffee

Teacher Support

Rebecca B. Corwin
Karen Economopoulos
Tracey Wright
Lisa Yaffee

Technology Development

Michael T. Battista
Douglas H. Clements
Julie Sarama Meredith
Andee Rubin

Video Production

David A. Smith

Administration and Production

Amy Catlin
Amy Taber

Cooperating Classrooms for This Unit

Michele de Silva
Boston Public Schools
Boston, MA

Alice Madio
Ambrose School
Winchester, MA

Consultants and Advisors

Elizabeth Badger
Deborah Lowenberg Ball
Marilyn Burns
Ann Grady
Joanne M. Gurry
James J. Kaput
Steven Leinwand
Mary M. Lindquist
David S. Moore
John Olive
Leslie P. Steffe
Peter Sullivan
Grayson Wheatley
Virginia Woolley
Anne Zarinnia

Graduate Assistants

Kent State University
Joanne Caniglia
Pam DeLong
Carol King

State University of New York at Buffalo
Rosa Gonzalez
Sue McMillen
Julie Sarama Meredith
Sudha Swaminathan

Revisions and Home Materials

Cathy Miles Grant
Marlene Kliman
Margaret McGaffigan
Megan Murray
Kim O'Neil
Andee Rubin
Susan Jo Russell
Lisa Seyferth
Myriam Steinback
Judy Storeygard
Anna Suarez
Cornelia Tierney
Carol Walker
Tracey Wright

CONTENTS

WHERE TO START

The first-time user of *Money, Miles, and Large Numbers* should read the following:

When you next teach this same unit, you can begin to read more of the background. Each time you present the unit, you will learn more about how your students understand the mathematical ideas.

Investigations in Number, Data, and Space® is a K–5 mathematics curriculum with four major goals:

- to offer students meaningful mathematical problems
- to emphasize depth in mathematical thinking rather than superficial exposure to a series of fragmented topics
- to communicate mathematics content and pedagogy to teachers
- to substantially expand the pool of mathematically literate students

The *Investigations* curriculum embodies a new approach based on years of research about how children learn mathematics. Each grade level consists of a set of separate units, each offering 2–8 weeks of work. These units of study are presented through investigations that involve students in the exploration of major mathematical ideas.

Approaching the mathematics content through investigations helps students develop flexibility and confidence in approaching problems, fluency in using mathematical skills and tools to solve problems, and proficiency in evaluating their solutions. Students also build a repertoire of ways to communicate about their mathematical thinking, while their enjoyment and appreciation of mathematics grows.

The investigations are carefully designed to invite all students into mathematics—girls and boys, members of diverse cultural, ethnic, and language groups, and students with different strengths and interests. Problem contexts often call on students to share experiences from their family, culture, or community. The curriculum eliminates barriers—such as work in isolation from peers, or emphasis on speed and memorization—that exclude some students from participating successfully in mathematics. The following aspects of the curriculum ensure that all students are included in significant mathematics learning:

- Students spend time exploring problems in depth.
- They find more than one solution to many of the problems they work on.

- They invent their own strategies and approaches, rather than relying on memorized procedures.
- They choose from a variety of concrete materials and appropriate technology, including calculators, as a natural part of their everyday mathematical work.
- They express their mathematical thinking through drawing, writing, and talking.
- They work in a variety of groupings—as a whole class, individually, in pairs, and in small groups.
- They move around the classroom as they explore the mathematics in their environment and talk with their peers.

While reading and other language activities are typically given a great deal of time and emphasis in elementary classrooms, mathematics often does not get the time it needs. If students are to experience mathematics in depth, they must have enough time to become engaged in real mathematical problems. We believe that a minimum of five hours of mathematics classroom time a week—about an hour a day—is critical at the elementary level. The plan and pacing of the *Investigations* curriculum is based on that belief.

We explain more about the pedagogy and principles that underlie these investigations in Teacher Notes throughout the units. For correlations of the curriculum to the NCTM Standards and further help in using this research-based program for teaching mathematics, see the following books:

- *Implementing the* Investigations in Number, Data, and Space® *Curriculum*
- *Beyond Arithmetic: Changing Mathematics in the Elementary Classroom* by Jan Mokros, Susan Jo Russell, and Karen Economopoulos

This book is one of the curriculum units for *Investigations in Number, Data, and Space.* In addition to providing part of a complete mathematics curriculum for your students, this unit offers information to support your own professional development. You, the teacher, are the person who will make this curriculum come alive in the classroom; the book for each unit is your main support system.

Although the curriculum does not include student textbooks, reproducible sheets for student work are provided in the unit and are also available as Student Activity Booklets. Students work actively with objects and experiences in their own environment and with a variety of manipulative materials and technology, rather than with a book of instruction and problems. We strongly recommend use of the overhead projector as a way to present problems, to focus group discussion, and to help students share ideas and strategies.

Ultimately, every teacher will use these investigations in ways that make sense for his or her particular style, the particular group of students, and the constraints and supports of a particular school environment. Each unit offers information and guidance for a wide variety of situations, drawn from our collaborations with many teachers and students over many years. Our goal in this book is to help you, a professional educator, implement this curriculum in a way that will give all your students access to mathematical power.

Investigation Format

The opening two pages of each investigation help you get ready for the work that follows.

What Happens This gives a synopsis of each session or block of sessions.

Mathematical Emphasis This lists the most important ideas and processes students will encounter in this investigation.

What to Plan Ahead of Time These lists alert you to materials to gather, sheets to duplicate, transparencies to make, and anything else you need to do before starting.

Everyday Uses of Money

What Happens

Sessions 1 and 2: Groceries, Lunch, and Book Orders Students estimate and find totals for items purchased at the grocery store. They share their strategies for finding estimates and exact answers. Students are introduced to and begin working on three Choice Time activities: Buying Groceries, What's for Lunch?, and First Grade Book Order.

Session 3: Making a Dollar Students identify combinations of two amounts and then three amounts that total $1.00 or 100. They play the card game Close to 100, which reinforces finding combinations close to 100. They may continue to play this game or work on other Choice Time activities from the previous sessions.

Sessions 4 and 5: Making Sense (Cents) of Money on the Calculator Students use mental estimation and calculators to solve addition problems involving money amounts. They discuss how to input various amounts of money and interpret the results as displayed on the calculators. They continue to work on choice activities, including a new one, Beat the Calculator. Session 5 ends with a discussion of the choice activity What's for Lunch?, in which students find the difference between two amounts.

Session 6: Making Change Students develop procedures for making change. They learn to count up as a way of making change. They do a number of transactions that involve making change.

Sessions 7 and 8: Shopping Smart Students finish working on their choice activities. They share their approaches for working on the first grade book order. They do an assessment task that involves estimating amounts, finding an exact total, and making change.

Mathematical Emphasis

- Estimating sums
- Estimating total amounts of money
- Exploring number relationships in the context of money
- Developing strategies for combining numbers, particularly money amounts
- Using landmark numbers (multiples of 10 or .10 and 100 or 1.00) to compare and find differences between two quantities
- Using standard addition and subtraction notation to record combining and comparing situations
- Using the calculator to solve problems
- Interpreting decimals on the calculator as amounts of money

What to Plan Ahead of Time

Materials

- Empty grocery cans and boxes (Sessions 1–3)
- Children's book catalogs (such as Trumpet Club, Scholastic) (Sessions 1–3)
- Calculators (1 per student)
- Coins for the overhead projector (Session 6, optional)
- Play coins and bills: some for each pair (Sessions 1–2; 6–8)
- Scissors
- Overhead projector (all sessions)

Other Preparation

- Duplicate student sheets and teaching resources, located at the end of this unit, in the following quantities. If you have Student Activity Booklets, copy only the items marked with an asterisk, including any transparencies needed.

For Sessions 1–2
Student Sheet 1, Different Ways—Same Amount (p. 63): 2 per student
Student Sheet 2, What's for Lunch? (p. 64): 2 per student
Student Sheet 3, First Grade Book Order (p. 65): 2 per student
Menus (pp. 75–76): 1 set per pair of students
Family letter* (p. 62): 1 per student. Remember to sign it before copying.
Student Sheet 4, What's in the Cupboard? (p. 66): 1 per student (homework)

For Session 3
Student Sheet 5, Close to 100 Score Sheet (p. 67): 3 per student, and 1 overhead transparency*
How to Play Close to 100 (p. 77): 1 per student

Numeral Cards (pp. 92–94): 1 deck per group and 1 deck per student for homework (see below)

For Sessions 4–5
Student Sheet 6, Money on the Calculator (p. 68): 1 per student (optional)
Student Sheet 7 (pages 1 and 2), Beat the Calculator (p. 69): 2 per student
Grocery Receipts* (p. 78): 1 transparency
Student Sheet 8, One Lunch Order (p. 71): 1 per student (homework)
Coins and Bills (p. 79): 1 per student, optional (homework)

For Session 6
Student Sheet 9, How Much Change? (p. 72): 1 per student (homework)
Coins and Bills (p. 79): 1 per student, optional (homework)

For Sessions 7–8
Student Sheet 10 (p. 73): Camping Supplies: 1 per student
Student Sheet 11, Art Supplies (p. 74): 1 per student (homework)

- Prior to teaching the unit, ask students to bring in empty (and clean) grocery cans and boxes. Add prices if needed.
- After Session 2, make transparencies of several students' lunch orders from the activity What's for Lunch? (for Session 5) and several students' book orders from the activity First Grade Book Order (for Session 7).
- If you haven't purchased the *Investigations* grade 4 Numeral Cards, make a deck of Numeral Cards for every two or three students. Cards duplicated on tagboard will last longer. Cut apart the 44 cards for each complete deck (students can help). Mark the back of each deck differently. (Session 3)

Sessions Within an investigation, the activities are organized by class session, a session being at least a one-hour math class. Sessions are numbered consecutively through an investigation. Often several sessions are grouped together, presenting a block of activities with a single major focus.

When you find a block of sessions presented together—for example, Sessions 1, 2, and 3—read through the entire block first to understand the overall flow and sequence of the activities. Make some preliminary decisions about how you will divide the activities into three sessions for your class, based on what you know about your students. You may need to modify your initial plans as you progress through the activities, and you may want to make notes in the margins of the pages as reminders for the next time you use the unit.

Be sure to read the Session Follow-Up section at the end of the session block to see what homework assignments and extensions are suggested as you make your initial plans.

While you may be used to a curriculum that tells you exactly what each class session should cover, we have found that the teacher is in a better position to make these decisions. Each unit is flexible and may be handled somewhat differently by every teacher. While we provide guidance for how many sessions a particular group of activities is likely to need, we want you to be active in determining an appropriate pace and the best transition points for your class. It is not unusual for a teacher to spend more or less time than is proposed for the activities.

Ten-Minute Math At the beginning of some sessions, you will find Ten-Minute Math activities. These are designed to be used in tandem with the investigations, but not during the math hour. Rather, we hope you will do them whenever you have a spare 10 minutes—maybe before lunch or recess, or at the end of the day.

Ten-Minute Math offers practice in key concepts, but not always those being covered in the unit. For example, in a unit on using data, Ten-Minute Math might revisit geometric activities done earlier in the year. Complete directions for the suggested activities are included at the end of each unit.

Sessions 1 and 2

Materials
- Overhead projector (optional)
- Calculators (1 per student)
- Play coins and bills (some for each pair)
- Empty cans, boxes, and cartons of grocery store items. If a price sticker is not on each item, make a new price sticker.
- Children's book catalogs (1 per pair)
- Student Sheet 1 (2 per student)
- Student Sheet 2 (2 per student)
- Student Sheet 3 (2 per student)
- Menus (1 set per pair)
- Family letter (1 per student)
- Student Sheet 4 (1 per student, homework)

Groceries, Lunch, and Book Orders

What Happens

Students estimate and find totals for items purchased at the grocery store. They share their strategies for finding estimates and exact answers. Students are introduced to and begin working on three Choice Time activities: Buying Groceries, What's for Lunch?, and First Grade Book Order. Their work focuses on:

- estimating totals for amounts of money
- adding money

Activity

Estimating and Finding Totals

On the board or overhead write:

$.99 $1.49

(Write these separately, not in a problem format, so students will develop their own ways to add them mentally.)

Suppose you are walking through the store and pick up two items. One costs $.99 and the other costs $1.49. You want to make sure you have enough money to pay for them. Since you don't have paper and pencil or a calculator handy, you need to figure mentally. How much money do you think you need?

Suggest to students that they write their estimates on paper when they are ready.

When most students have recorded their estimates, have them share their answers and how they arrived at them. Look especially for solutions that take into account the value of the numbers by using landmarks. Here are some possible strategies:

4 ■ *Investigation 1: Everyday Uses of Money*

Activities The activities include pair and small-group work, individual tasks, and whole-class discussions. In any case, students are seated together, talking and sharing ideas during all work times. Students most often work cooperatively, although each student may record work individually.

Choice Time In some units, some sessions are structured with activity choices. In these cases, students may work simultaneously on different activities focused on the same mathematical ideas. Students choose which activities they want to do, and they cycle through them.

You will need to decide how to set up and introduce these activities and how to let students make their choices. Some teachers present them as station activities, in different parts of the room. Some list the choices on the board as reminders or have students keep their own lists.

Extensions Sometimes in Session Follow-Up, you will find suggested extension activities. These are opportunities for some or all students to explore a topic in greater depth or in a different context. They are not designed for "fast" students; mathematics is a multifaceted discipline, and different students will want to go further in different investigations. Look for and encourage the sparks of interest and enthusiasm you see in your students, and use the extensions to help them pursue these interests.

Excursions Some of the *Investigations* units include excursions—blocks of activities that could be omitted without harming the integrity of the unit. This is one way of dealing with the great depth and variety of elementary mathematics— much more than a class has time to explore in any one year. Excursions give you the flexibility to make different choices from year to year, doing the excursion in one unit this time, and next year trying another excursion.

Tips for the Linguistically Diverse Classroom At strategic points in each unit, you will find concrete suggestions for simple modifications of the teaching strategies to encourage the participation of all students. Many of these tips offer alternative ways to elicit critical thinking from students at varying levels of English proficiency, as well as from other students who find it difficult to verbalize their thinking.

The tips are supported by suggestions for specific vocabulary work to help ensure that all students can participate fully in the investigations. The Preview for the Linguistically Diverse Classroom (p. I-21) lists important words that are assumed as part of the working vocabulary of the unit. Second-language learners will need to become familiar with these words in order to understand the problems and activities they will be doing. These terms can be incorporated into students' second-language work before or during the unit. Activities that can be used to present the words are found in the appendix, Vocabulary Support for Second-Language Learners (p. 60). In addition, ideas for making connections to students' language and cultures, included on the Preview page, help the class explore the unit's concepts from a multicultural perspective.

Materials

A complete list of the materials needed for teaching this unit is found on p. I-16. Some of these materials are available in kits for the *Investigations* curriculum. Individual items can also be purchased from school supply dealers.

Classroom Materials In an active mathematics classroom, certain basic materials should be available at all times: interlocking cubes, pencils, unlined paper, graph paper, calculators, things to count with, and measuring tools. Some activities in this curriculum require scissors and glue sticks or tape. Stick-on notes and large paper are also useful materials throughout.

So that students can independently get what they need at any time, they should know where these materials are kept, how they are stored, and how they are to be returned to the storage area. For example, interlocking cubes are best stored in towers of ten; then, whatever the activity, they should be returned to storage in groups of ten at the end of the hour. You'll find that establishing such routines at the beginning of the year is well worth the time and effort.

Technology Calculators are used throughout *Investigations.* Many of the units recommend that you have at least one calculator for each pair. You will find calculator activities, plus Teacher Notes discussing this important mathematical tool, in an early unit at each grade level. It is assumed that calculators will be readily available for student use.

Computer activities at grade 4 use a software program that was developed especially for the *Investigations* curriculum. The program *Geo-Logo™* is used for activities in the 2-D Geometry unit, *Sunken Ships and Grid Patterns,* where students explore coordinate graphing systems, the use of negative numbers to represent locations in space, and the properties of geometric figures.

How you use the computer activities depends on the number of computers you have available. Suggestions are offered in the geometry units for how to organize different types of computer environments.

Children's Literature Each unit offers a list of suggested children's literature (p. I-16) that can be used to support the mathematical ideas in the unit. Sometimes an activity is based on a specific children's book, with suggestions for substitutions where practical. While such activities can be adapted and taught without the book, the literature offers a rich introduction and should be used whenever possible.

Student Sheets and Teaching Resources Student recording sheets and other teaching tools needed for both class and homework are provided as reproducible blackline masters at the end of each unit. They are also available as Student Activity Booklets. These booklets contain all the sheets each student will need for individual work, freeing you from extensive copying (although you may need or want to copy the occasional teaching resource on transparency film or card stock, or make extra copies of a student sheet).

We think it's important that students find their own ways of organizing and recording their work. They need to learn how to explain their thinking with both drawings and written words, and how to organize their results so someone else can under-

stand them. For this reason, we deliberately do not provide student sheets for every activity. Regardless of the form in which students do their work, we recommend that they keep a mathematics notebook or folder so that their work is always available for reference.

Homework In *Investigations,* homework is an extension of classroom work. Sometimes it offers review and practice of work done in class, sometimes preparation for upcoming activities, and sometimes numerical practice that revisits work in earlier units. Homework plays a role both in supporting students' learning and in helping inform families about the ways in which students in this curriculum work with mathematical ideas.

Depending on your school's homework policies and your own judgment, you may want to assign more homework than is suggested in the units. For this purpose you might use the practice pages, included as blackline masters at the end of this unit, to give students additional work with numbers.

Name _____ Date _____

Student Sheet 1

Different Ways—Same Amount

Choose two grocery items. Record what they are and what their prices are. Use play money to show the cost in two different ways. Record both ways, drawing the coins (and bills) below.

Switch items with your partner. On a new student sheet, find two ways of showing prices for your partner's items.

Compare your results. Have you and your partner found the same ways or different ways? Have you or your partner found the way that uses the fewest coins?

1. One item I chose is: _____

   Its price is: _____

   Two ways of showing the price are:

2. Another item I chose is: _____

   Its price is: _____

   Two ways of showing the price are:

© Dale Seymour Publications®   63   *Investigation 1 • Sessions 1–2*
Money, Miles, and Large Numbers

For some homework assignments, you will want to adapt the activity to meet the needs of a variety of students in your class: those with special needs, those ready for more challenge, and second-language learners. You might change the numbers in a problem, make the activity more or less complex, or go through a sample activity with those who need extra help. You can modify any student sheet for either homework or class use. In particular, making numbers in a problem smaller or larger can make the same basic activity appropriate for a wider range of students.

Another issue to consider is how to handle the homework that students bring back to class—how to recognize the work they have done at home without spending too much time on it. Some teachers hold a short group discussion of different approaches to the assignment; others ask students to share and discuss their work with a neighbor, or post the homework around the room and give students time to tour it briefly. If you want to keep track of homework students bring in, be sure it ends up in a designated place.

Investigations at Home It is a good idea to make your policy on homework explicit to both students and their families when you begin teaching with *Investigations*. How frequently will you be assigning homework? When do you expect homework to be completed and brought back to school? What are your goals in assigning homework? How independent should families expect their children to be? What should the parent's or guardian's role be? The more explicit you can be about your expectations, the better the homework experience will be for everyone.

Investigations at Home (a booklet available separately for each unit, to send home with students) gives you a way to communicate with families about the work students are doing in class. This booklet includes a brief description of every session, a list of the mathematics content emphasized in each investigation, and a discussion of each homework assignment to help families more effectively support their children. Whether or not you are using the *Investigations* at Home booklets, we expect you to make your own choices about home-

work assignments. Feel free to omit any and to add extra ones you think are appropriate.

Family Letter A letter that you can send home to students' families is included with the blackline masters for each unit. Families need to be informed about the mathematics work in your classroom; they should be encouraged to participate in and support their children's work. A reminder to send home the letter for each unit appears in one of the early investigations. These letters are also available separately in Spanish, Vietnamese, Cantonese, Hmong, and Cambodian.

Help for You, the Teacher

Because we believe strongly that a new curriculum must help teachers think in new ways about mathematics and about their students' mathematical thinking processes, we have included a great deal of material to help you learn more about both.

About the Mathematics in This Unit This introductory section (p. I-17) summarizes the critical information about the mathematics you will be teaching. It describes the unit's central mathematical ideas and how students will encounter them through the unit's activities.

Teacher Notes These reference notes provide practical information about the mathematics you are teaching and about our experience with how students learn. Many of the notes were written in response to actual questions from teachers, or to discuss important things we saw happening in the field-test classrooms. Some teachers like to read them all before starting the unit, then review them as they come up in particular investigations.

Dialogue Boxes Sample dialogues demonstrate how students typically express their mathematical ideas, what issues and confusions arise in their thinking, and how some teachers have guided class discussions.

These dialogues are based on the extensive classroom testing of this curriculum; many are word-for-word transcriptions of recorded class discussions. They are not always easy reading; sometimes it may take some effort to unravel what the students are trying to say. But this is the value of these dialogues; they offer good clues to how your students may develop and express their approaches and strategies, helping you prepare for your own class discussions.

Where to Start You may not have time to read everything the first time you use this unit. As a first-time user, you will likely focus on understanding the activities and working them out with your students. Read completely through each investigation before starting to present it. Also read those sections listed in the Contents under the heading Where to Start (p. vi).

How to Use This Book ■ **I-7**

The *Investigations* curriculum incorporates the use of two forms of technology in the classroom: calculators and computers. Calculators are assumed to be standard classroom materials, available for student use in any unit. Computers are explicitly linked to one or more units at each grade level; they are used with the unit on 2-D geometry at each grade, as well as with some of the units on measuring, data, and changes.

Using Calculators

In this curriculum, calculators are considered tools for doing mathematics, similar to pattern blocks or interlocking cubes. Just as with other tools, students must learn both *how* to use calculators correctly and *when* they are appropriate to use. This knowledge is crucial for daily life, as calculators are now a standard way of handling numerical operations, both at work and at home.

Using a calculator correctly is not a simple task; it depends on a good knowledge of the four operations and of the number system, so that students can select suitable calculations and also determine what a reasonable result would be. These skills are the basis of any work with numbers, whether or not a calculator is involved.

Unfortunately, calculators are often seen as tools to check computations with, as if other methods are somehow more fallible. Students need to understand that any computational method can be used to check any other; it's just as easy to make a mistake on the calculator as it is to make a mistake on paper or with mental arithmetic. Throughout this curriculum, we encourage students to solve computation problems in more than one way in order to double-check their accuracy. We present mental arithmetic, paper-and-pencil computation, and calculators as three possible approaches.

In this curriculum we also recognize that, despite their importance, calculators are not always appropriate in mathematics instruction. Like any tools, calculators are useful for some tasks, but not for others. You will need to make decisions about when to allow students access to calculators and when to ask that they solve problems without them, so that they can concentrate on other tools and skills. At times when calculators are or are not appropriate for a particular activity, we make specific recommendations. Help your students develop their own sense of which problems they can tackle with their own reasoning and which ones might be better solved with a combination of their own reasoning and the calculator.

Managing calculators in your classroom so that they are a tool, and not a distraction, requires some planning. When calculators are first introduced, students often want to use them for everything, even problems that can be solved quite simply by other methods. However, once the novelty wears off, students are just as interested in developing their own strategies, especially when these strategies are emphasized and valued in the classroom. Over time, students will come to recognize the ease and value of solving problems mentally, with paper and pencil, or with manipulatives, while also understanding the power of the calculator to facilitate work with larger numbers.

Experience shows that if calculators are available only occasionally, students become excited and distracted when they are permitted to use them. They focus on the tool rather than on the mathematics. In order to learn when calculators are appropriate and when they are not, students must have easy access to them and use them routinely in their work.

If you have a calculator for each student, and if you think your students can accept the responsibility, you might allow them to keep their calculators with the rest of their individual materials, at least for the first few weeks of school. Alternatively, you might store them in boxes on a shelf, number each calculator, and assign a corresponding number to each student. This system can give students a sense of ownership while also helping you keep track of the calculators.

Using Computers

Students can use computers to approach and visualize mathematical situations in new ways. The computer allows students to construct and manipulate geometric shapes, see objects move according to rules they specify, and turn, flip, and repeat a pattern.

This curriculum calls for computers in units where they are a particularly effective tool for learning mathematics content. One unit on 2-D geometry at each of the grades 3–5 includes a core of activities that rely on access to computers, either in the classroom or in a lab. Other units on geometry, measurement, data, and changes include computer activities, but can be taught without them. In these units, however, students' experience is greatly enhanced by computer use.

The following list outlines the recommended use of computers in this curriculum:

Grade 1
Unit: *Survey Questions and Secret Rules*
   (Collecting and Sorting Data)
Software: *Tabletop, Jr.*
Source: Broderbund

Unit: *Quilt Squares and Block Towns*
   (2-D and 3-D Geometry)
Software: *Shapes*
Source: provided with the unit

Grade 2
Unit: *Mathematical Thinking at Grade 2*
   (Introduction)
Software: *Shapes*
Source: provided with the unit

Unit: *Shapes, Halves, and Symmetry*
   (Geometry and Fractions)
Software: *Shapes*
Source: provided with the unit

Unit: *How Long? How Far?* (Measuring)
Software: *Geo-Logo*
Source: provided with the unit

Grade 3
Unit: *Flips, Turns, and Area* (2-D Geometry)
Software: *Tumbling Tetrominoes*
Source: provided with the unit

Unit: *Turtle Paths* (2-D Geometry)
Software: *Geo-Logo*
Source: provided with the unit

Grade 4
Unit: *Sunken Ships and Grid Patterns*
   (2-D Geometry)
Software: *Geo-Logo*
Source: provided with the unit

Grade 5
Unit: *Picturing Polygons* (2-D Geometry)
Software: *Geo-Logo*
Source: provided with the unit

Unit: *Patterns of Change* (Tables and Graphs)
Software: *Trips*
Source: provided with the unit

Unit: *Data: Kids, Cats, and Ads* (Statistics)
Software: Tabletop, Sr.
Source: Broderbund

The software provided with the *Investigations* units uses the power of the computer to help students explore mathematical ideas and relationships that cannot be explored in the same way with physical materials. With the *Shapes* (grades 1–2) and *Tumbling Tetrominoes* (grade 3) software, students explore symmetry, pattern, rotation and reflection, area, and characteristics of 2-D shapes. With the *Geo-Logo* software (grades 3–5), students investigate rotations and reflections, coordinate geometry, the properties of 2-D shapes, and angles. The *Trips* software (grade 5) is a mathematical exploration of motion in which students run experiments and interpret data presented in graphs and tables.

We suggest that students work in pairs on the computer; this not only maximizes computer resources but also encourages students to consult, monitor, and teach one another. Generally, more than two students at one computer find it difficult to share. Managing access to computers is an issue for every classroom. The curriculum gives you explicit support for setting up a system. The units are structured on the assumption that you have enough computers for half your students to work on the machines in pairs at one time. If you do not have access to that many computers, suggestions are made for structuring class time to use the unit with five to eight computers, or even with fewer than five.

Assessment plays a critical role in teaching and learning, and it is an integral part of the *Investigations* curriculum. For a teacher using these units, assessment is an ongoing process. You observe students' discussions and explanations of their strategies on a daily basis and examine their work as it evolves. While students are busy recording and representing their work, working on projects, sharing with partners, and playing mathematical games, you have many opportunities to observe their mathematical thinking. What you learn through observation guides your decisions about how to proceed. In any of the units, you will repeatedly consider questions like these:

■ Do students come up with their own strategies for solving problems, or do they expect others to tell them what to do? What do their strategies reveal about their mathematical understanding?

■ Do students understand that there are different strategies for solving problems? Do they articulate their strategies and try to understand other students' strategies?

■ How effectively do students use materials as tools to help with their mathematical work?

■ Do students have effective ideas for keeping track of and recording their work? Does keeping track of and recording their work seem difficult for them?

You will need to develop a comfortable and efficient system for recording and keeping track of your observations. Some teachers keep a clipboard handy and jot notes on a class list or on adhesive labels that are later transferred to student files. Others keep loose-leaf notebooks with a page for each student and make weekly notes about what they have observed in class.

Assessment Tools in the Unit

With the activities in each unit, you will find questions to guide your thinking while observing the students at work. You will also find two built-in assessment tools: Teacher Checkpoints and embedded Assessment activities.

Teacher Checkpoints The designated Teacher Checkpoints in each unit offer a time to "check in" with individual students, watch them at work, and ask questions that illuminate how they are thinking.

At first it may be hard to know what to look for, hard to know what kinds of questions to ask. Students may be reluctant to talk; they may not be accustomed to having the teacher ask them about their work, or they may not know how to explain their thinking. Two important ingredients of this process are asking students open-ended questions about their work and showing genuine interest in how they are approaching the task. When students see that you are interested in their thinking and are counting on them to come up with their own ways of solving problems, they may surprise you with the depth of their understanding.

Teacher Checkpoints also give you the chance to pause in the teaching sequence and reflect on how your class is doing overall. Think about whether you need to adjust your pacing: Are most students fluent with strategies for solving a particular kind of problem? Are they just starting to formulate good strategies? Or are they still struggling with how to start? Depending on what you see as the students work, you may want to spend more time on similar problems, change some of the problems to use smaller numbers, move quickly to more challenging material, modify subsequent activities for some students, work on particular ideas with a small group, or pair students who have good strategies with those who are having more difficulty.

Embedded Assessment Activities Assessment activities embedded in each unit will help you examine specific pieces of student work, figure out what it means, and provide feedback. From the students' point of view, these assessment activities are no different from any others. Each is a learning experience in and of itself, as well as an opportunity for you to gather evidence about students' mathematical understanding.

The embedded assessment activities sometimes involve writing and reflecting; at other times, a discussion or brief interaction between student and teacher; and in still other instances, the creation and explanation of a product. In most cases, the assessments require that students *show* what they did, *write* or *talk* about it, or do both. Having to explain how they worked through a problem helps students be more focused and clear in their mathematical thinking. It also helps them realize that doing mathematics is a process that may involve tentative starts, revising one's approach, taking different paths, and working through ideas.

Teachers often find the hardest part of assessment to be interpreting their students' work. We provide guidelines to help with that interpretation. If you have used a process approach to teaching writing, the assessment in *Investigations* will seem familiar. For many of the assessment activities, a Teacher Note provides examples of student work and a commentary on what it indicates about student thinking.

Documentation of Student Growth

To form an overall picture of mathematical progress, it is important to document each student's work in journals, notebooks, or portfolios. The choice is largely a matter of personal preference; some teachers have students keep a notebook or folder for each unit, while others prefer one mathematics notebook, or a portfolio of selected work for the entire year. The final activity in each *Investigations* unit, called Choosing Student Work to Save, helps you and the students select representative samples for a record of their work.

This kind of regular documentation helps you synthesize information about each student as a mathematical learner. From different pieces of evidence, you can put together the big picture. This synthesis will be invaluable in thinking about where to go next with a particular child, deciding where more work is needed, or explaining to parents (or other teachers) how a child is doing.

If you use portfolios, you need to collect a good balance of work, yet avoid being swamped with an overwhelming amount of paper. Following are some tips for effective portfolios:

■ Collect a representative sample of work, including some pieces that students themselves select for inclusion in the portfolio. There should be just a few pieces for each unit, showing different kinds of work—some assignments that involve writing, as well as some that do not.

■ If students do not date their work, do so yourself so that you can reconstruct the order in which pieces were done.

■ Include your reflections on the work. When you are looking back over the whole year, such comments are reminders of what seemed especially interesting about a particular piece; they can also be helpful to other teachers and to parents. Older students should be encouraged to write their own reflections about their work.

Assessment Overview

There are two places to turn for a preview of the assessment opportunities in each *Investigations* unit. The Assessment Resources column in the unit Overview Chart (pp. I-13–I-15) identifies the Teacher Checkpoints and Assessment activities embedded in each investigation, guidelines for observing the students that appear within classroom activities, and any Teacher Notes and Dialogue Boxes that explain what to look for and what types of student responses you might expect to see in your classroom. Additionally, the section About the Assessment in This Unit (p. I-19) gives you a detailed list of questions for each investigation, keyed to the mathematical emphases, to help you observe student growth.

Depending on your situation, you may want to provide additional assessment opportunities. Most of the investigations lend themselves to more frequent assessment, simply by having students do more writing and recording while they are working.

Money, Miles, and Large Numbers

Content of This Unit   This unit involves students adding and subtracting decimal numbers and numbers in the hundreds and thousands. Students work with combining situations in which they determine how several numbers combine to make a given number, and comparing situations that involve finding the difference between two quantities. They do these in the context of money and distance, which are natural situations where decimals and larger numbers occur. For example, they recommend $100.00 worth of children's books for the school library; they estimate and measure distances around their school and neighborhood using miles and tenths of miles; and they plan a trip around the United States keeping track of miles traveled.

Emphasis throughout the unit is placed on students developing their own sound strategies for adding and subtracting. Students are encouraged to use estimation and multiple strategies to double-check their work. The unit also includes number games and problems that build students' fluency with using 100 and 1000 and multiples of 100 and 1000 as important landmarks in combining and comparing. The use of landmark numbers is extended to include decimals, with students using numbers such as $1.00 and $10.00 as landmarks for money amounts.

Connections with Other Units   If you are doing the full-year *Investigations* curriculum in the suggested sequence for grade 4, this is the seventh of eleven units. Some of the activities in this unit are direct extensions of the grade 4 units *Mathematical Thinking at Grade 4* and *Landmarks in the Thousands*. These prior units lay the groundwork for effective combining and comparing strategies.

This unit can be used successfully at either grade 4 or grade 5, depending on the previous experience and needs of your students.

Investigations Curriculum   ■   Suggested Grade 4 Sequence

Mathematical Thinking at Grade 4 (Introduction)

Arrays and Shares (Multiplication and Division)

Seeing Solids and Silhouettes (3-D Geometry)

Landmarks in the Thousands (The Number System)

Different Shapes, Equal Pieces (Fractions and Area)

The Shape of the Data (Statistics)

▶*Money, Miles, and Large Numbers* (Addition and Subtraction)

Changes Over Time (Graphs)

Packages and Groups (Multiplication and Division)

Sunken Ships and Grid Patterns (2-D Geometry)

Three out of Four Like Spaghetti (Data and Fractions)

Investigation 1 ▪ Everyday Uses of Money

Class Sessions	Activities	Pacing
Sessions 1 and 2 (p. 4) GROCERIES, LUNCH, AND BOOK ORDERS	Estimating and Finding Totals Choice Time: Groceries, Menus, and Book Orders Teacher Checkpoint: Choice Time Homework: What's in the Cupboard?	minimum 2 hr
Session 3 (p. 13) MAKING A DOLLAR	Ways to Make a Dollar Choice Time Activities Homework: Close to 100 Extension: Ways to Make 100	minimum 1 hr
Sessions 4 and 5 (p. 18) MAKING SENSE (CENTS) OF MONEY ON THE CALCULATOR	Adding Money on the Calculator: What Happens to the Cents? Estimating Grocery Receipts Beat the Calculator and Other Choice Time Activities Class Discussion: What's for Lunch? Homework: One Lunch Order Extension: Shopping at Home	minimum 2 hr
Session 6 (p. 23) MAKING CHANGE	Change from $1.00 Teacher Checkpoint: Spaghetti and Sauce Homework: How Much Change? Extension: Exact Change	minimum 1 hr
Sessions 7 and 8 (p. 27) SHOPPING SMART	Completing Choice Time Activities Class Discussion: First Grade Book Orders Assessment: Camping Supplies Homework: Art Supplies	minimum 2 hr

◔ **Ten-Minute Math** ▪ **Likely or Unlikely?**

Mathematical Emphasis

- Exploring number relationships in the context of money

- Developing strategies for combining numbers, particularly money amounts

- Using landmark numbers (multiples of 10 or .10 and 100 or 1.00) to compare and find differences between two quantities

- Using standard addition and subtraction notation to record combining and comparing situations

- Using the calculator to solve problems

- Interpreting decimals on the calculator as amounts of money

Assessment Resources

Teacher Checkpoint: Choice Time (p. 8)

Using Landmark Numbers for Comparing (Teacher Note, p. 9)

Three Powerful Addition Strategies (Teacher Note, p. 10)

How Much Money Would You Need? (Dialogue Box, p. 12)

Playing Close to 100 (Teacher Note, p. 16)

Teacher Checkpoint: Spaghetti and Sauce (p. 24)

Keeping Track of Addition and Subtraction (Teacher Note, p. 26)

Assessment: Camping Supplies (p. 28)

Materials

Empty grocery cans and boxes

Children's book catalogs

Calculators

Coins for display

Play coins and bills

Scissors

Overhead projector

Student Sheets 1–11

Family letter

Investigation 2 ▪ How Far? Measuring in Miles and Tenths

Class Sessions	Activities	Pacing
Sessions 1 and 2 (p. 32) MILES AND TENTHS OF A MILE	Introducing Parts of a Mile Running Distances Teacher Checkpoint: Making Your Own Log Calculator Problems	minimum 2 hr
Session 3 (Excursion)* (p. 36) HOW FAR IS $1/10$ OF A MILE?	How Far is $1/10$ of a Mile? Homework: My Measuring Method	minimum 1 hr
Session 4 (p. 41) A TOUR OF OUR TOWN	City Landmarks Planning the Tour Homework: My Tour	minimum 1 hr

*Excursions can be omitted without harming the integrity or continuity of the unit, but they offer good mathematical work if you have time to include them.

Mathematical Emphasis

- Estimating local distances in miles and tenths of miles; developing a sense of approximate length of a mile and $1/10$ of a mile

- Comparing and combining decimal numbers and finding differences between these numbers

- Seeing the relationships of decimal parts to the whole

- Measuring distances on maps using a scale

- Becoming familiar with common decimal and fraction equivalents

- Estimating and calculating sums of quantities that include decimal portions

Assessment Resources

Observing the Students (p. 33)

Teacher Checkpoint: Making Your Own Log (p. 34)

Measuring $1/10$ of a Mile (Dialogue Box, p. 39)

How Far Is It to the Post Office? (Dialogue Box, p. 45)

Materials

Calculators

String

Adding machine tape

Yardsticks

Scissors

Map of your city, town, or neighborhood

Index cards

Rulers

Student Sheets 12–18

Investigation 3 ▪ Calculating Longer Distances

Class Sessions	Activities	Pacing
Session 1 (p. 48) CLOSE TO 1000	Playing Close to 1000 Class Discussion: Strategies for Adding and Subtracting Large Numbers Homework: Close to 1000	minimum 1 hr
Sessions 2, 3, and 4 (p. 51) A TRIP AROUND THE UNITED STATES	Measuring Distance on a Map A Trip Around the United States Assessment: Observing Students and Analyzing Their Travel Logs Choosing Student Work to Save Homework: A Trip Around the United States Extension: Famous Journeys	minimum 3 hr

◗ **Ten-Minute Math** ▪ **Likely or Unlikely?**

Mathematical Emphasis

- Measuring distances on maps using a scale

- Comparing and combining numbers in the hundreds and thousands

- Using standard addition and subtraction notation to record combining and comparing problems

Assessment Resources

Assessment: Observing Students and Analyzing Their Travel Logs (p. 54)

Choosing Student Work to Save (p. 55)

Materials

Calculators

Tape measures

Wall map of the U.S.

Strips of paper (optional)

Inch cubes (optional)

Travel guides or brochures for the U.S.

Overhead projector (optional)

Student Sheets 19–21

Teaching resource sheets

Following are the basic materials needed for the activities in this unit. Many of the items can be purchased from the publisher, either individually or in the Teacher Resource Package and the Student Materials Kit for grade 4. Detailed information is available on the *Investigations* order form. To obtain this form, call toll-free 1-800-872-1100 and ask for a Dale Seymour customer service representative.

Numeral Cards (manufactured; or use blackline masters to make your own sets—1 deck per group for class work, 2 decks per student for homework)

Empty cans, boxes, and cartons from the grocery store

String (3–4 rolls)

Adding machine tape (3–4 rolls)

Calculators (1 per student)

Children's book catalogs (for example, Trumpet Club, Scholastic)

Index cards (1 per student)

Maps of your city, town, or neighborhood with a scale of miles (2 per group)

Play coins and bills (optional)

Rulers: 1 per pair

Scissors

Coins for the overhead projector (optional)

Overhead projector

Tape measure (1 per pair)

Travel guides or brochures for the United States (2–3 per pair)

Wall maps of the United States

Yardsticks (3–4)

Strips of paper (optional)

Inch cubes (optional)

The following materials are provided at the end of this unit as blackline masters. A Student Activity Booklet containing all student sheets and teaching resources needed for individual work is available.

Family Letter (p. 62)

Student Sheets 1–21 (p. 63)

Teaching Resources:

    Menus (pp. 75–76)

    How to Play Close to 100 (p. 77)

    Grocery Receipts (p. 78)

    Coins and Bills (p. 79)

    How to Play Close to 1000 (p. 91)

    Numeral Cards (pp. 92–94)

Practice Pages (p. 95)

Related Children's Literature

Axelrod, Amy. *Pigs Will Be Pigs*. New York: Four Winds Press, 1994.

Burningham, John. *Around the World in Eighty Days*. London: Jonathan Cape, 1972.

Milton, Nancy. *The Giraffe That Walked to Paris*. New York: Crown, 1992.

Schwartz, David. *How Much Is a Million?* New York: Lothrop, Lee and Shepard, 1985.

Schwartz, David. *If You Made a Million*. New York: Lothrop, Lee and Shepard, 1989.

This unit is about addition and subtraction in contexts using decimal numbers and numbers in the hundred and thousands. The emphasis throughout the unit is on combining and comparing numbers. The investigations are not based on straightforward, immediately obvious addition or subtraction situations, but instead they require students to think about what's actually involved in the problem context: "If I am planning a trip around the United States and cannot travel more than 10,000 miles, how many more miles can I travel if I have already gone 4234 miles? Is this an addition or a subtraction problem? Should I add on to or subtract from 10,000?"

How would you solve it? How can it be represented using standard addition or subtraction notation? A key goal of this unit is for students to learn *how* to use addition and subtraction flexibly to solve problems.

Students' own strategies—strategies that involve good mental arithmetic, estimation, and a sound understanding of number relationships—are what matter here. Students must develop combining and comparing strategies that make sense to them. For example, when estimating grocery receipts ($2.88, $1.29, and $1.77) students may use a combination of addition and subtraction to make a close estimate: "$2.88 is almost $3.00, $1.77 is almost $2.00; that's about $5.00, and another $1.00 makes it about $6.00. I won't add on the 29 cents because I added extra to the other numbers." If they wanted to figure an exact total they might continue: "I added extra to the $1.77, and $2.88, that's 23 cents plus 12 cents, that's 35 extra cents, so I'll subtract the 29 cents on $1.29 from that. That's still 6 cents too much, so I have to subtract 6 cents from the $6.00. It's $5.94 in all."

This strategy is based on sound understanding of the relationship between numbers and some key landmarks in the numbers system—in this case, multiples of 100. One of the most important goals of the unit is to encourage students to articulate, develop, and use their own strategies for solving comparing and combining problems. As they develop and share these strategies with their peers, they are also encouraged to apply them to many different situations. As students make calcu-

lations, they must also create systems for keeping track of their work. In the above problem, many students would jot down some intermediate steps to keep track of their procedures.

Throughout the unit, students have many opportunities to work with numbers as they appear in real-world contexts: money and measuring distance. By seeing numbers as they naturally appear, such as seeing decimals as parts of familiar things like a dollar and a mile, they become accustomed to looking for them and interpreting them in their environment outside of school.

During this unit, students will be thinking a great deal about the relationships between numbers or quantities without relying on memorized procedures. While there is nothing wrong with knowing the computational procedures that have traditionally been taught in American schools, this unit has a different emphasis. Too often, students memorize procedures they don't understand, then apply them blindly without engaging in any mathematical thought. When emphasis is on learning rote procedures, many students just try to follow the steps and seem to forget everything they know about relationships among numbers. They don't estimate, they don't think about the situation, and they don't check their results. Rather, we want students to develop procedures they can rely on because they are based on understandings about numbers and number relationships. For example, problems like the following are often particularly difficult for students who use the standard subtraction algorithm, even for students who are adept at these procedures:

$$\begin{array}{r} 1001 \\ -3 \\ \hline \end{array}$$

The regrouping procedure typically taught in school is unnecessarily complex for this computation. In fact, for the child who understands how these numbers relate to each other, who sees 1000 as a meaningful landmark in the number system, and who has a sense of the relationship between these two numbers, this problem is quite easy.

One of your most important tasks as a teacher during this unit will be to make sure that students use only approaches they can understand and explain.

It may take some work before they believe you value their thinking and expect them to use what they know about number relationships to reason about addition and subtraction situations.

Mathematical Emphasis At the beginning of each investigation, the Mathematical Emphasis section tells you what is most important for students to learn about during that investigation. Many of these mathematical understandings and processes are difficult and complex. Students gradually learn more and more about each idea over many years of schooling.

Individual students will begin and end the unit with different levels of knowledge and skill, but all students will develop strategies that make sense to them when adding and subtracting, learn to apply these strategies to problem situations, and recognize and read standard notation that describes these addition and subtraction situations. Students will also learn to keep track of their strategies through making notes and writing down intermediate steps, to estimate to check for reasonableness, and to double-check their work by using more than one approach to solve a problem.

Throughout the *Investigations* curriculum, there are many opportunities for ongoing daily assessment as you observe, listen to, and interact with students at work. In this unit, you will find three Teacher Checkpoints:

Investigation 1, Sessions 1–2:
Choice Time (p. 8)

Investigation 1, Session 6:
Spaghetti and Sauce (p. 24)

Investigation 2, Sessions 1–2:
Making Your Own Log (p. 34)

This unit also has two embedded assessment activities:

Investigation 1, Sessions 7–8:
Camping Supplies (p. 28)

Investigation 3, Sessions 2–4:
Observing Students and Analyzing Their Travel Logs (p. 54)

In addition, you can use almost any activity in this unit to assess your students' needs and strengths. Listed below are questions to help you focus your observation in each investigation. You may want to keep track of your observations for each student to help you plan your curriculum and monitor students' growth. Suggestions for documenting student growth can be found in the section About Assessment (p. I–10).

Investigation 1: Everyday Uses of Money

■ What number relationships do students notice and use to solve money problems? Do they use landmarks? Equivalencies?

■ How do students make an estimate for a problem that involves combining several numbers or amounts of money? Are their estimates reasonable? How do they solve such problems exactly? Do they estimate and then adjust? Do they work mentally? With pencil and paper? With manipulatives? How do students use manipulatives, such as coins, bills, and calculators?

■ What landmarks do students use to compare numbers? How do students use landmarks to find the difference between two quantities? What strategies do students use to make change?

■ How do students record and keep track of their work? How accurate and comfortable are they using addition and subtraction notation? Can students communicate clearly about the way they solved the problem?

■ How do students use a calculator to solve a problem? Do they analyze the results to see if they are reasonable?

■ How do students use calculators to add and subtract decimals? How do they interpret decimal numbers as amounts of money? Can they enter an amount of money on the calculator? Can they interpret a number on the display as an amount of money?

Investigation 2: How Far? Measuring in Miles and Tenths

■ How do students estimate distances? What strategies and information do they use? Are they developing a better sense of a mile and $1/10$ of a mile?

■ How do students compare and combine decimal numbers? How do they find the difference between two decimal numbers? What strategies do they use? Do they use landmarks? Manipulatives? Calculators? Do they count on (or back) from one number? Do they consider whether their answers are reasonable?

■ How do students relate a part of a decimal (such as $1/10$ of a mile) to the whole mile? How do students make sense of and represent these relationships? How do they keep track of their work?

■ How accurately do students measure the distance between two places on a map? How do they make sense of and use a scale to measure distance on a map? Do they use repeated addition? Landmarks? Multiplication?

■ How do students estimate and calculate exact sums when working with quantities that include decimal portions? What strategies and information do they use? Do they estimate and then adjust? Do they use landmarks? Manipulatives? Common decimal and fraction equivalents ($1/4$ of a mile is the same as .25 mile)?

Investigation 3: Calculating
Longer Distances

- How accurately do students measure the distance between two places on a map? How do they make sense of and use a scale to measure distance on a map? Do they use repeated addition? Landmarks? Multiplication? Are students able to plan trips of no more than 10,000 miles with between 4 and 10 stops? How do they represent their trips on a map of the United States?

- What strategies do students use to combine and compare numbers in the hundreds and thousands? Do they estimate and then adjust? Do they use landmarks? Do they consider the largest portion of a number first, moving from left to right? Do they count up (or down) from one number to another to find the distance? How flexible are they in using such strategies?

- How do students interpret problems presented with traditional addition and subtraction notation? Do they use a strategy that is appropriate for the problem? How do they use addition and subtraction notation to represent their work?

In the *Investigations* curriculum, mathematical vocabulary is introduced naturally during the activities. We don't ask students to learn definitions of new terms; rather, they come to understand such words as *factor, area,* and *symmetry* by hearing them used frequently in discussion as they investigate new concepts. This approach is compatible with current theories of second-language acquisition, which emphasize the use of new vocabulary in meaningful contexts while students are actively involved with objects, pictures, and physical movement.

Listed below are some key words used in this unit that will not be new to most English speakers at this age level but may be unfamiliar to students with limited English proficiency. You will want to spend additional time working on these words with your students who are learning English. If your students are working with a second-language teacher, you might enlist your colleague's aid in familiarizing students with these words before and during this unit. In the classroom, look for opportunities for students to hear and use these words. Activities you can use to present the words are given in the appendix, Vocabulary Support for Second-Language Learners (p. 60).

menu, restaurant, order   Using restaurant menus, students make up lunch orders that cost not more than $6.00. They choose at least three items from the menus, record their orders, total the cost, and then determine the change they would receive from $6.00.

receipts, prices, clerk, grocery store, change   Students estimate and find the exact total of grocery receipts. In pairs, they assume the role of customer and clerk, one paying for the groceries and the other calculating the amount of change due back. Both students are responsible for double-checking the calculations.

mileage, miles, odometer   Students consider distances measured in miles and tenths of miles. They compare odometer readings recorded by their teacher and determine how far the car traveled. Students also engage in two activities, a runner's log and a trip around the United States, where they combine and compare distances recorded in miles.

Multicultural Extensions for All Students
Whenever possible, encourage students to share words, objects, customs, or any aspects of daily life from their own cultures and backgrounds that are relevant to the activities in this unit. For example:

- Encourage students to include on their grocery lists foods from shops where they or their families purchase foods from their countries of origin.

- Students could choose to create new menus that include foods from their countries of origin and cultures.

- As an extension to the United States trip, have students plan a trip where they can visit different countries. Students could plan a trip through their native countries, highlighting places they would visit.

Investigations

INVESTIGATION 1

Everyday Uses of Money

What Happens

Sessions 1 and 2: Groceries, Lunch, and Book Orders Students estimate and find totals for items purchased at the grocery store. They share their strategies for finding estimates and exact answers. Students are introduced to and begin working on three Choice Time activities: Buying Groceries, What's for Lunch?, and First Grade Book Order.

Session 3: Making a Dollar Students identify combinations of two amounts and then three amounts that total $1.00 or 100. They play the card game Close to 100, which reinforces finding combinations close to 100. They may continue to play this game or work on other Choice Time activities from the previous sessions.

Sessions 4 and 5: Making Sense (Cents) of Money on the Calculator Students use mental estimation and calculators to solve addition problems involving money amounts. They discuss how to input various amounts of money and interpret the results as displayed on the calculators. They continue to work on choice activities, including a new one, Beat the Calculator. Session 5 ends with a discussion of the choice activity What's for Lunch?, in which students find the difference between two amounts.

Session 6: Making Change Students develop procedures for making change. They learn to count up as a way of making change. They do a number of transactions that involve making change.

Sessions 7 and 8: Shopping Smart Students finish working on their choice activities. They share their approaches for working on the first grade book order. They do an assessment task that involves estimating amounts, finding an exact total, and making change.

Mathematical Emphasis

- Estimating sums
- Estimating total amounts of money
- Exploring number relationships in the context of money
- Developing strategies for combining numbers, particularly money amounts
- Using landmark numbers (multiples of 10 or .10 and 100 or 1.00) to compare and find differences between two quantities
- Using standard addition and subtraction notation to record combining and comparing situations
- Using the calculator to solve problems
- Interpreting decimals on the calculator as amounts of money

What to Plan Ahead of Time

Materials

- Empty grocery cans and boxes (Sessions 1–3)
- Children's book catalogs (such as Trumpet Club, Scholastic) (Sessions 1–3)
- Calculators (1 per student)
- Coins for the overhead projector (Session 6, optional)
- Play coins and bills: some for each pair (Sessions 1–2; 6–8)
- Scissors
- Overhead projector (all sessions)

Other Preparation

- Duplicate student sheets and teaching resources, located at the end of this unit, in the following quantities. If you have Student Activity Booklets, copy only the items marked with an asterisk, including any transparencies needed.

  For Sessions 1–2

  Student Sheet 1, Different Ways—Same Amount (p. 63): 2 per student

  Student Sheet 2, What's for Lunch? (p. 64): 2 per student

  Student Sheet 3, First Grade Book Order (p. 65): 2 per student

  Menus (pp. 75–76): 1 set per pair of students

  Family letter* (p. 62): 1 per student. Remember to sign it before copying.

  Student Sheet 4, What's in the Cupboard? (p. 66): 1 per student (homework)

  For Session 3

  Student Sheet 5, Close to 100 Score Sheet (p. 67): 3 per student, and 1 overhead transparency*

  How to Play Close to 100 (p. 77): 1 per student

  Numeral Cards (pp. 92–94): 1 deck per group and 1 deck per student for homework (see below)

  For Sessions 4–5

  Student Sheet 6, Money on the Calculator (p. 68): 1 per student (optional)

  Student Sheet 7 (pages 1 and 2), Beat the Calculator (p. 69): 2 per student

  Grocery Receipts* (p. 78): 1 transparency

  Student Sheet 8, One Lunch Order (p. 71): 1 per student (homework)

  Coins and Bills (p. 79): 1 per student, optional (homework)

  For Session 6

  Student Sheet 9, How Much Change? (p. 72): 1 per student (homework)

  Coins and Bills (p. 79): 1 per student, optional (homework)

  For Sessions 7–8

  Student Sheet 10 (p. 73): Camping Supplies: 1 per student

  Student Sheet 11, Art Supplies (p. 74): 1 per student (homework)

- Prior to teaching the unit, ask students to bring in empty (and clean) grocery cans and boxes. Add prices if needed.
- After Session 2, make transparencies of several students' lunch orders from the activity What's for Lunch? (for Session 5) and several students' book orders from the activity First Grade Book Order (for Session 7).
- If you haven't purchased the *Investigations* grade 4 Numeral Cards, make a deck of Numeral Cards for every two or three students. Cards duplicated on tagboard will last longer. Cut apart the 44 cards for each complete deck (students can help). Mark the back of each deck differently. (Session 3)

Materials

- Overhead projector (optional)
- Calculators (1 per student)
- Play coins and bills (some for each pair)
- Empty cans, boxes, and cartons of grocery store items. If a price sticker is not on each item, make a new price sticker.
- Children's book catalogs (1 per pair)
- Student Sheet 1 (2 per student)
- Student Sheet 2 (2 per student)
- Student Sheet 3 (2 per student)
- Menus (1 set per pair)
- Family letter (1 per student)
- Student Sheet 4 (1 per student, homework)

Groceries, Lunch, and Book Orders

What Happens

Students estimate and find totals for items purchased at the grocery store. They share their strategies for finding estimates and exact answers. Students are introduced to and begin working on three Choice Time activities: Buying Groceries, What's for Lunch?, and First Grade Book Order. Their work focuses on:

- estimating totals for amounts of money
- adding money

Activity

Estimating and Finding Totals

On the board or overhead write:

$ .99 $1.49

(Write these separately, not in a problem format, so students will develop their own ways to add them mentally.)

Suppose you are walking through the store and pick up two items. One costs $.99 and the other costs $1.49. You want to make sure you have enough money to pay for them. Since you don't have paper and pencil or a calculator handy, you need to figure mentally. How much money do you think you need?

Suggest to students that they write their estimates on paper when they are ready.

When most students have recorded their estimates, have them share their answers and how they arrived at them. Look especially for solutions that take into account the value of the numbers by using landmarks. Here are some possible strategies:

"I thought 99 cents was close to $1.00 and $1.49 was close to $1.50, so that's $2.50."

"It's $2.48, since $.99 is 1¢ less than $1.00, and $1.49 is 1¢ less than $1.50, so 2¢ less than $2.50 is $2.48."

Accept both estimates and exact calculations. Ask students to distinguish between which answers are estimates and which are exact calculations. Tell students that an estimate is close enough for many purposes, such as knowing whether you have enough money to pay for things you buy. See the **Teacher Note**, Using Landmark Numbers for Comparing (p. 9).

Put another set of prices on the board or overhead:

$1.79   74¢

How much money do you need to buy these two items? First find a quick estimate.

Again, students record their estimates then share their strategies. These are some possible responses from fourth graders:

$1.79 is about $2.00, and 74¢ is almost $1.00, so that would be $3.00.

$1.79 is about $2.00 and 74¢ more would be $2.74.

$1.79 is about $1.80, and 74¢ is about 75¢, so it would be $2.75 minus 20¢, since $1.80 is 20¢ less than $2.00, or about $2.55.

Now see if you can find an exact answer to this problem without using paper and pencil.

Students spend a few minutes mentally calculating their answers. Record all student answers on the board or overhead before asking them to explain their strategies for calculating. (In this way, students won't forget answers while others are explaining their thinking.)

Repeat this procedure for another set of prices: $2.59 and $1.84, and, as an extension, $2.88, $1.29, and 77¢. Ask students to find a quick estimate and then to calculate mentally an exact answer. Then have students share their strategies for finding estimates and exact answers. See the **Dialogue Box**: How Much Money Would You Need? (p. 12).

Don't expect all your students to have found the correct answers to these problems. At this point, you want to emphasize that there are many ways of finding the answers, and that students need to listen to one another and see if they can understand others' approaches. See the **Teacher Note**, Three Powerful Addition Strategies (p. 10).

Choice Time: Groceries, Menus, and Book Orders

Introducing Choice Time   Throughout this investigation, students will choose from a variety of activities that are going on simultaneously in the classroom. This format allows students to explore the same idea at different paces. If you have done other units in the *Investigations* curriculum, students may be familiar with the Choice Time format. Spend some time talking with your students about how they make their choices, where the materials are, and where they should work.

How to Set Up the Choices   For the remainder of this session and the next, students are engaged in Choice Time. If you set up your choices at stations, show students what they will find at each station. Otherwise, tell students where they can find the materials they need.

> Choice 1: Buying Groceries—copies of Student Sheet 1; play money; empty cans, boxes, or cartons from the grocery store with prices (original or made up) marked on them
>
> Choice 2: What's for Lunch?—copies of Student Sheet 2, menus
>
> Choice 3: First Grade Book Order—copies of Student Sheet 3, children's book catalogs

Students will need to keep track of the choices that are available and the ones they have completed. You may want to have students make lists of the activities they do on blank sheets of paper. If you are having students keep track of their work in individual folders, distribute these when you introduce choices.

Calculators should be available for all sessions of this investigation.

❖   **Tip for the Linguistically Diverse Classroom**   Pair limited English proficient students with English proficient students; have proficient students read the Student Sheets aloud to their partners.

Choice 1: Buying Groceries
There are two activities within this choice. Students should work with partners. They will need to have five or six grocery store containers and some play money.

A. Different Ways—Same Amount   Each person picks two of the containers. They record on Student Sheet 1, Different Ways—Same Amount, what they are and the price of the first item. They use coins and bills to show the cost in two different ways. They do the same for the second item. Partners then exchange items, and using a new student sheet, each represents the prices of her or his partner's items in two ways. They compare their results, noting similar and different ways of representing the cost. Finally, students try to find the way that uses the fewest number of coins.

B. Making Change   One person plays the customer and chooses one thing to buy. This person pays for the item with either a $1 bill or a $5 bill. The other person is the clerk and makes change using the play money. Then they switch roles so each has an opportunity to be the customer and the clerk. When they are finished, they write a short description of the transaction. For example: "When I was the customer, I bought a carton of milk that cost $1.49. I gave the clerk $5.00. I got back $3.51 in change."

Choice 2: What's for Lunch?

Students have $6.00 to spend for lunch. They choose and order from one of the four restaurant menus provided. They write their lunch order of at least three items on Student Sheet 2, What's for Lunch? They total the cost, then find the change they would receive from $6.00. After they have completed their order from one restaurant, they may want to make an order from another restaurant.

Choice 3: First Grade Book Order

Using a catalog of children's books, students recommend books for a first grader. They have $100.00 to spend. They write down the titles and prices on the order form (Student Sheet 3). Remind them they have only $100.00 and should choose a variety of books they think first graders would like.

After you have introduced each activity, explain to students that they will be working on these activities during the next week.

In this investigation we will be doing a number of choice activities that involve addition and subtraction of money. You may do them in any order, and you don't have to finish all of the choices in one day. You may want to do some of the activities several times. The choice is up to you, but by the end of the unit, I hope you will have completed most of the activities.

Activity

Teacher Checkpoint
Choice Time

Observing the Students

During Choice Time, circulate around the room and observe students as they are working:

- How are they combining numbers?
- What strategies do they use to make change?
- Are they using estimates as part of their procedure?
- Do they do the problems mentally? With pencil and paper? With a calculator?
- How are they keeping track of their work?

As you observe students, ask each to explain his or her strategy for solving a problem if it is not readily apparent in their work.

While students are working independently, you may want to meet with small groups to help those having difficulty or to learn about the different strategies students are using to solve these money problems.

Remind students they will have more opportunities to work on these choice activities in the following sessions and do not need to complete all three activities.

Note: As students complete the two choice activities involving lunch and book orders, make transparencies of several of their orders for Sessions 5 and 7.

Sessions 1 and 2 Follow-Up

Homework

What's in the Cupboard? Students choose an amount such as $3.00, $5.00, or $10.00 and then find grocery items in their cupboards at home with prices that total around that amount. They make a list of these items, their prices, and their total cost on Student Sheet 4, What's in the Cupboard? If students do not have access to groceries with price labels, they could use items from the drugstore or hardware store. Stress that the total need not be exact; it should just be close to $3.00, $5.00, or $10.00. At the beginning of Session 2, have a few students record their lists on the board or overhead, without the total cost, and have students estimate the total cost. Send home the family letter or *Investigations* at Home booklet today.

Using Landmark Numbers for Comparing

Number sense involves a deep understanding of numbers, their characteristics, their place in the number system, and their relationships to one another. Think for a minute about this problem: "If you are 48 years old and I am 62, how much older am I than you?"

When you think about finding the difference between 48 and 62, you immediately bring your number sense into play. You recognize very quickly a great deal about these numbers. You might use any of a number of ideas—62 is 2 more than 60, 48 is 2 less than 50, 48 + 10 is 58, 62 – 10 is 52, and 10 is the difference between 50 and 60.

A lot of the information we use to solve this kind of problem has to do with the relationship between a quantity in the problem and a nearby landmark in the number system. For example, you could solve this problem by thinking that 48 to 50 is 2, 50 to 60 is 10, 60 to 62 is 2 more, so the difference is 10 + 2 + 2, or 14. You used multiples of ten—50 and 60—as critical landmarks or anchors to which the numbers 48 and 62 are connected.

One purpose of this unit is for students to become more fluent with the use of such land-marks in the number system to solve addition and subtraction problems. As their work moves into numbers in the hundreds, multiples of 100 will join multiples of 10 as critical landmarks.

In order to use these critical numbers effectively, students must become fluent in finding the difference between a number they are using and nearby landmarks. The game Close to 100 provides practice in this skill. ("If I have 48, how much more do I need to make 100? If I have made a sum of 87, how far away is that from 100?") As students play this game, encourage them to use what they know about number relationships. If a student is trying to figure out what is needed to add to 48 to get 100, he or she might count by 10's: "58, 68, 78, 88, 98—that's five 10's, and 2 more is 52." Another student might use 50 as a landmark: "I know 50 to 100 is 50, and 48 to 50 is 2, so it's 52."

Powerful mental arithmetic strategies are based on this kind of knowledge. As students become more fluent in using what they know about number relationships and critical landmarks in the number system, they will develop strategies they can rely on to solve addition and subtraction problems.

Most of us who are teaching today learned to add starting with the ones, then the tens, then the hundreds, and so on, moving from right to left and "carrying" from one column to another. This algorithm is certainly efficient once it's mastered. However, there are many other ways of adding that are just as efficient, that are closer to how we naturally think about quantities, that connect better with good estimation strategies, and that generally result in fewer errors.

When students rely only on memorized rules and procedures they do not understand, they usually do not estimate or double-check. They make mistakes that make no sense, considering the numbers. We want students to use strategies that encourage, rather than discourage, them to think about the quantities they are using and what to expect as the result. We want them to use their knowledge of the number system and important landmarks in that system. We want them to easily break apart and recombine numbers in ways that help them make computation more straightforward and therefore less prone to error.

The three powerful addition strategies discussed here are familiar to many competent users of mathematics. Your students may well invent others. It is critical that every student be comfortable with more than one way of adding so an answer obtained using one method can be checked by using another. Anyone can make a mistake while doing routine computation—even with a calculator. What is critical, when accuracy matters, is that you have spent enough time estimating and double-checking to be able to rely on your result.

Left-to-Right Addition: Biggest Quantities First When students develop their own strategies for addition from an early age, they usually move from left to right, starting with the bigger parts of the quantities. For example, when adding 47 + 48, a student might say "40 and 40 is 80, then 8 and 7 is 15, so 80 plus 10 more is 90, then 5 more makes 95." This strategy is both efficient and accurate. Some people who are extremely good at computation use this strategy

as their basic approach to addition, even with large numbers.

One advantage of this approach is that when students work with the largest quantities first, it's easier for them to maintain a good sense of what the final sum should be. Another advantage is that students keep seeing the quantities 47 and 48 as whole quantities, rather than breaking them up into their separate digits and losing track of the whole. When using the traditional algorithm (8 + 7 = 15, put down the 5, carry the 1), students too often see the 8, the 7, the 5, the 1, and the two 4's as individual digits. They lose their sense of the quantities involved, and if they end up with a nonsensical answer, they do not see it because they "did it the right way."

Using Nearby Landmarks Changing an unfamiliar number to a more familiar one that is easier to compute is another strategy students should develop. Multiples of 10 and multiples of 100 are especially useful landmarks for students at this age. For example, in order to add 199 and 149, you might think of the problem as 200 plus 150, find the total of 350, then subtract 2 to compensate for the 2 added on at the beginning.

Of course, there are other useful landmarks, too. If you are adding 23, 26, and 27, you might use 25 as your landmark, rather than 20 or 30: "Three 25's would be 75, so I'm 2 under and 2 over with the 23 and 27; I just add 1 more from the 26, and it's 76."

There are no rules about which landmarks in the number system are best. It simply depends on whether using landmarks helps you solve the problem.

Changing the Order of the Numbers Simply changing the order of the numbers you are adding is often a great help. For example, when adding 23 + 46 + 7, the problem becomes much simpler as soon as you recognize that 23 + 7 is 30. Changing the order of numbers can also involve partitioning some numbers into two

Continued on next page

parts. For example, if you are adding 108 + 45 +162, you might add this way: "160 plus 40 is 200, plus another 100 is 300; 2 and 8 is 10, plus 5 is 15, so it's 315."

These strategies may be used alone or in combination, whether the problem is being done mentally, on paper, or with the calculator. Encourage students to get into the habit of always looking over the whole problem before they begin solving it. Are there numbers they can combine easily? Are there useful landmark numbers they can use? Will they solve it by adding from left to right?

There are no hard-and-fast rules about which strategies are best for which problems. It really depends on what works for a particular person and how that person sees a particular problem. Even when you may think that a particular strategy is clearly best, students can surprise you in the way that they see the problem. For example,

we may think it is obvious to change 199 + 149 to 200 + 150, then subtract 2. However, someone else might use the following method, just as efficient and accurate: "199 plus 100 is 299, then I'll take 1 from the 49 to make 300, leaving 48, so it's 348." Similarly, while one person might use 25 as a landmark to solve 23 + 26 + 27, another might rearrange the numbers in the problem, adding 23 + 27 first to get 50, then adding on the 26.

If you have students who have already memorized the traditional right-to-left algorithm and believe this is how they are "supposed" to do addition, you will have to work hard to instill some new values—that estimating the result is critical, having more than one strategy is a necessary part of doing computation, and using what you know about the numbers to simplify the problem leads to procedures that make more sense.

How Much Money Would You Need?

Students discuss their strategies for estimating the sum of $2.88, $1.29, and 77¢ in the activity Estimating and Finding Totals (p. 4).

Sarah: I think you would need about $6.00.

Why do you say $6.00?

Sarah: Because you'd need at least $3.00 for the $2.88, and $1.00 for the 77 cents, and $2.00 for the $1.29.

What does anyone else think? Do you think about it the same way or differently?

Alex: I think you should say $5.00 because with the $1.29, you'd round it down to a dollar.

So would $5.00 be enough? What do other people think?

Rafael: But I think it was better what Sarah did, because if you make it more, then you'd be sure to have enough.

Marci: I think $5.00 would be enough because for the $2.88 you went up to $3.00, and for the 77 cents you went up to $1.00, so you already have extra money; it doesn't matter if you go down for the $1.29.

Other opinions?

Luisa: I did it differently. I said $1.29 was $1.30, $2.88 is about $2.90, and 77 cents is near 80 cents. If we made the $2.90 into $3.00 and the 80 cents into a whole dollar, we'd need 30 cents more. We can give the 30 cents from the $1.30 to them, and everything will be dollars: one and three and one. That's $5.00.

Do you think it would be over or under $5.00?

Luisa: A little bit under.

Rafael: I think it would be pretty close to $5.00, too.

Why?

Rafael: Because on some you're going up, and on some you're going down.

Making a Dollar

What Happens

Students identify combinations of two amounts and then three amounts that total $1.00 or 100. They play the card game Close to 100, which reinforces finding combinations close to 100. They may continue to play this game or work on other Choice Time activities from the previous sessions. Their work focuses on:

■ finding combinations to total $1.00 or 100

  Ten-Minute Math: Likely or Unlikely?   Once or twice in the next few days, do Likely or Unlikely? with your students. This activity does not need to happen during math class but can be done during any free ten minutes of the day.

The first time you do this activity, you will need to prepare some likely/unlikely statements ahead of time. In the future you can have students write likely/unlikely statements as they arrive in the morning or for homework.

List the headings Likely and Unlikely on the board.

Read each statement and have students categorize them as either likely or unlikely.

Keep the list posted in the classroom so it can be added to each time you do this activity.

For full directions and variations, see p. 58.

Materials

■ Choice Time materials from Session 2

■ Numeral Cards (1 deck per group for class work, 1 deck per student for homework)

■ Student Sheet 5 (2 per student, 1 for homework)

■ Transparency of Student Sheet 5 (optional)

■ How to Play Close to 100 (1 per student)

■ Overhead projector

Activity

Ways to Make a Dollar

Discuss various combinations that total one dollar:

Yesterday I went to the store. I bought two things that together cost exactly one dollar. What could have been the cost of each of the two things?

Record a few students' responses on the board—such as 50¢ and 50¢; 30¢ and 70¢. Then ask students to record different combinations of the cost of two items that total one dollar.

On the board or overhead, make a list of combinations students came up with. Have them share their strategies for figuring out combinations totaling $1.00. Did they make a random list, or are their combinations related? Some students might use the strategy of adding one cent to one amount and subtracting one cent from the second amount (for example, 99¢ + 1¢,

98¢ + 2¢, and so on). Others may add 10 cents to one amount and subtract 10 cents from the other (for example, 33¢ + 67¢, 43¢ + 57¢, and so on).

What if I went to the store and bought three things that totaled a dollar? What could have been the cost of each of the three things?

Again, record a few initial responses on the board. Then have students, working alone or in pairs, record many different combinations of three prices that total one dollar.

When you observe students working on this problem, ask how they are finding new combinations. Do they have a strategy that helps them get many combinations easily? Does anyone have a strategy that changes all three prices each time?

Activity

Choice Time Activities

Students spend the remainder of this session engaged in Choice Time activities. Introduce them to playing Close to 100, or they can continue working on the other choice activities, Buying Groceries, What's for Lunch?, and First Grade Book Order. Make sure all of the choice materials are available.

Close to 100 Close to 100 focuses on finding pairs of two-digit numbers that make a total as close to 100 as possible. This game was introduced in the unit, *Mathematical Thinking at Grade 4*, and the game Close to 1000 was suggested as a Choice Time activity in *Landmarks in the Thousands*. Read the instructions for Close to 100 (p. 77) and Close to 1000 (p. 91), to decide which version of the game is appropriate for your students at this time. Also read the **Teacher Note**, Playing Close to 100 (p. 16), for examples of how students might play the game. Close to 100 is added as a choice in the next few sessions, and Close to 1000 is played in the next investigation to work with larger numbers.

Introduce students to the game Close to 100 or review it by playing a demonstration game with the class. Give each student a copy of How to Play Close to 100 (p. 77).

Display six Numeral Cards and ask everyone to write two two-digit numbers on slips of paper or cards. All students try to construct the two numbers so that their sum is as close as possible to 100. Students share results, and scores are calculated for each result. Each player's score is the difference of the sum of the two numbers from 100.

Show students, perhaps using a transparency of Student Sheet 5, how to score the game. If your students have played the game before, teach them the alternative way to score with positive and negative numbers. If they are playing the game for the first time today, save this scoring alternative to introduce to students playing the game during Choice Time.

Scoring Close to 100 with + and –   This alternative scoring variation is described in the instructions for How to Play Close to 100 and Close to 1000. In this version, students score +3 if they are 3 over 100 and –3 if they are 3 under 100. This scoring changes players' strategies. The object is now to balance the scores over 100 with the scores under 100 to get as close to 0 as possible. Don't expect students to understand how the scoring makes a difference to their strategy right away. Tell them to play a practice game so they can see how the scoring works. Students who understand the idea of positive and negative changes can also use the calculator to help them total their score. Once you have introduced the variation, students can split into groups of two or three to play the game. If you wait to introduce this scoring during Choice Time, introduce the new scoring to six or eight students at a time. Students who have learned the new version can teach other students.

Session 3 Follow-Up

Close to 100   Students teach Close to 100 to someone at home who hasn't played before. They will need a deck of Numeral Cards, a copy of Student Sheet 5, and the How to Play Close to 100 instructions.

Ways to Make 100   Some students may be interested in investigating all the combinations of two whole numbers that total 100. When they think they have found them all, they write a convincing statement about how many combinations there are and how they know.

The Basic Game

Here is a sample of two students playing Close to 100, using the basic scoring.

Round 1—

Alex is dealt:	5	8	6	9	2	7
Lina Li is dealt:	9	1	5	5	4	7

Alex makes 58 + 29.   Lina Li makes 45 + 57.

Round 2—

Alex has 6 and 7 left from Round 1 and is dealt: 3 6 9 2

Lina Li has 9 and 1 left from Round 1 and is dealt: 8 2 5 0

Alex makes 36 + 62. Lina Li makes 98 + 02.

Note: Alex could have gotten closer to 100 in Round 2. Can you see how?

Alex's complete game went like this:

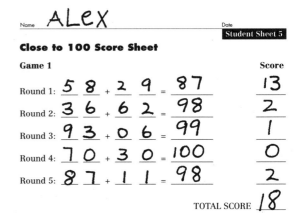

Name ALEX   Date
                                              Student Sheet 5

Close to 100 Score Sheet

Game 1			Score
Round 1:	5 8 + 2 9 =	87	13
Round 2:	3 6 + 6 2 =	98	2
Round 3:	9 3 + 0 6 =	99	1
Round 4:	7 0 + 3 0 =	100	0
Round 5:	8 7 + 1 1 =	98	2

TOTAL SCORE 18

Lina Li's complete game went like this:

Name Lina Li   Date
                                              Student Sheet 5

Close to 100 Score Sheet

Game 1			Score
Round 1:	4 5 + 5 7 =	102	2
Round 2:	9 8 + 0 2 =	100	0
Round 3:	6 2 + 5 1 =	113	13
Round 4:	4 7 + 4 9 =	96	4
Round 5:	8 5 + 0 6 =	91	9

TOTAL SCORE 28

Continued on next page

Playing with Negative and Positive Integers

Note: Students should be very comfortable with the basic game before trying this variation.

In this variation, students score the game Close to 100 using negative and positive integers. If you score 103, the difference from 100 is +3, so that is your score. If it is 98, the difference from 100 is –2, so that is your score. So, for example, the score sheets from the sample game on page 30 would look like these sample score sheets:

Name **ALEX** Date
Student Sheet 5

Close to 100 Score Sheet

Game 1   Score

Round 1: 5 8 + 2 9 = 87   –13
Round 2: 3 6 + 6 2 = 98   –2
Round 3: 9 3 + 0 6 = 99   –1
Round 4: 7 0 + 3 0 = 100   0
Round 5: 8 7 + 1 1 = 98   –2

TOTAL SCORE –18

Lina Li's complete game went like this:

Name **Lina Li** Date
Student Sheet 5

Close to 100 Score Sheet

Game 1   Score

Round 1: 4 5 + 5 7 = 102   +2
Round 2: 9 8 + 0 2 = 100   0
Round 3: 6 2 + 5 1 = 113   +13
Round 4: 4 7 + 4 9 = 96   –4
Round 5: 8 5 + 0 6 = 91   –9

TOTAL SCORE +2

The player with the total score *closest to 0* wins. In this case, –18 is 18 away from 0, and +2 is 2 away from 0, so Lina Li wins.

Scoring this way changes the strategy for the game. Even though Alex got four scores very close to 100, he did not compensate for his negative values with some positive ones. Lina Li had some totals further away from 100, but she balanced her negative and positive scores more evenly to come out with a total score closer to zero.

Making Sense (Cents) of Money on the Calculator

Materials

- Choice Time materials
- Calculators (1 per student)
- Student Sheet 6 (1 per student, optional)
- Student Sheet 7 (1 per student)
- Transparency of Grocery Receipts
- Menus from Sessions 1–2 (1 set per pair)
- Transparency of several students' lunch orders
- Overhead projector
- Student Sheet 8 (1 per student, homework)
- Coins and Bills (1 per student, homework,

What Happens

Students use mental estimation and calculators to solve addition problems involving money amounts. They discuss how to input various amounts of money and interpret the results as displayed on the calculators. They continue to work on choice activities, including a new one, Beat the Calculator. Session 5 ends with a discussion of the choice activity What's for Lunch?, in which students find the difference between two amounts. Their work focuses on:

- inputting different amounts of money on the calculator
- interpreting calculator displays as amounts of money
- adding money on the calculator

Activity

Adding Money on the Calculator: What Happens to the Cents?

Write the following amounts on the board or overhead (not in the form of an addition problem), and ask students to find the total of these amounts using their calculators.

$4.83 $24.39 $56.28

What number is displayed on the calculator? Why doesn't the calculator show $85.50?

Explain to students that many calculators drop extra zeros on the right, past the decimal point. Show that .5 and .50 both mean one half because .5 is 5 out of 10 and .50 is 50 out of 100. On the board write:

$.50 $.5   .50   50¢

Just as $.50 means half a dollar or 50 cents, .50 is one way of writing one half. The calculator doesn't put a zero at the end of decimals—it shows one half the simplest way, by displaying just .5.

We're going to do some simple problems on your calculators, problems you would usually do in your head, so we can think about how calculators show amounts of money. Do this problem mentally first. How much is $1.25 and 2 cents?

Even though you know the answer, what would you enter on your calculator to do this problem? Pay close attention to what buttons you are using to enter this information so you can share what you did.

Students report exactly what they entered on their calculator, such as, "One, decimal point, two, five, plus decimal point, zero, two, equals." Record responses as they are stated (1.25 + .02 =). Check to see if all students know they need to enter .02 on the calculator to show two cents.

Here is another problem to do in your head. How much is 40 cents and 20 cents? What would you enter on your calculator to do this problem?

Again have students tell you exactly what they entered on their calculators to do this problem. If students reply "40 + 20 =" (without using decimals), ask them how they would do the problem using decimals to show cents. Establish that you would enter ".40 + .20 =" on the calculator.

Some calculators show 0.4 when you enter .40. Why do you suppose that is? Why is there a zero in front?

What answer is shown on the calculator after you have added 20 cents? Why does zero point six (0.6) mean sixty cents?

Someone told me that if I wanted to add 40 cents and 20 cents on the calculator, I could enter ".4 + .2 =" and get the correct answer—that I don't need to enter the zeros to the right of the 4 and 2. Is this true? Why?

If students see the equivalency of .4 and .40, they may find that omitting the zeros can be a shortcut for them. However, this can be confusing for many students. If they seem confused about which zeros can be omitted and which are necessary to enter, urge them to continue entering money on the calculator in the same way they would write the amounts on paper.

How comfortable students feel with this activity will depend on how much experience they have using calculators. If your students need practice entering and interpreting amounts of money on the calculator, have them do Student Sheet 6, Money on the Calculator. They can work in pairs and, when they are finished, discuss two or three of the problems with the whole group.

If you feel your students are comfortable with entering amounts of money on the calculator, Student Sheet 6 can be an optional activity.

Estimating Grocery Receipts

Yesterday I stopped at the grocery store for just a few things. As I was taking the items out of my basket and putting them on the moving belt at the check-out counter, I mentally counted and added the prices in my head. When the clerk totaled the amount on the cash register, I found out that the total in my head was very close to the amount on the cash register—it was within $1.00.

Here is a list of groceries with their prices. [*Show the top half of the transparency of Grocery Receipts.*] **Estimate the total and then I'll show you how I estimated the total**.

Have students share their estimates and the strategies they used. Students may want to demonstrate their strategies on the overhead projector. Solicit as many different approaches as you can. If none of your students used the "counting dollars method" (counting dollars and combining cents to make a dollar), share the following approach. (If it's similar to a strategy used by a student, point out the similarities.)

Cereal	2.39
Tuna	.62
Bread	1.47
Milk	.75
Ice Cream	2.47
Cheese	.78

Here's the method I used to do a quick estimate—I counted dollars and combined cents to make dollars. I counted $2.00 for the cereal and added the dollars of the bread, $1.00, and the ice cream, $2.00, to make $5.00. Then I combined the cents of the cereal and tuna, which was about $1.00, and that brought the total up to around $6.00. The combined cents of the bread and ice cream was about $1.00, which made $7.00. I added $1.50 more for the milk and cheese, so the total was about $8.50.

As you share your strategies, you may want to highlight the amounts you are working with by circling them or grouping them in some way.

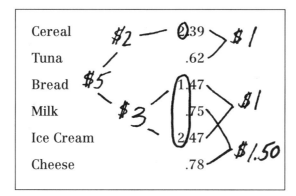

Ask the students to check the exact total using calculators. Compare the exact total ($8.48) with the estimates.

Show the students the items on the bottom half of the transparency of Grocery Receipts. Ask them to make a quick estimate and share their strategies. If no one uses a counting dollars approach, ask them to think about how they might count whole dollars to estimate this total. See the **Teacher Note**, Three Powerful Addition Strategies (p. 10).

Activity

Beat the Calculator and Other Choice Time Activities

Introduce students to a new Choice Time activity called Beat the Calculator, which involves estimating grocery receipts. Working in pairs, students determine whether it is faster to mentally estimate the total of grocery receipts, within a dollar, or to calculate the total with a calculator. For Beat the Calculator, each student needs a copy of Student Sheet 7 (pages 1 and 2), Beat the Calculator, and each pair needs a calculator. Students take turns estimating the total and using the calculator to find an exact total.

For the remainder of this session and the next, students continue working on the choices: Buying Groceries, What's for Lunch?, First Grade Book Order, Close to 100, and Beat the Calculator. Plan to have a discussion about the choice What's for Lunch? at the end of Session 5. In preparation for this discussion, make an overhead transparency of two or three students' lunch orders (Student Sheet 2) to use as part of the discussion.

Activity

Class Discussion: What's for Lunch?

Students will need to have copies of the menus to refer to. On the overhead projector, display a copy of a student's three-item lunch order. Show only the three items; cover up the total.

I have made transparencies of several of your orders. This is Emilio's lunch order. What restaurant did he go to? He ordered three things. What is the total cost of these things?

Collect all answers. If there are any differences, have students check their answers.

How much change did Emilio receive from $6.00?

Have students share their strategies for mentally calculating the change.

Repeat this activity for two or three other students' lunch orders. Vary the questions to students. Choose an order where the student ordered more than three items.

As students share their estimates and strategies, listen for evidence that suggests they are using an effective method of making a quick and reasonable estimate and that they are able to calculate totals accurately. As they share their strategies for figuring out how much change they would get back from $6.00, take note of how they are calculating this amount: Are students subtracting the total from $6.00? Are they counting on from the total to $6.00? Either strategy is appropriate as long as it makes sense to the student and he or she can explain the method used. In Session 6, students will further develop their strategies for making change.

Sessions 4 and 5 Follow-Up

 **Homework**

One Lunch Order   For homework, students estimate and then calculate the total amount spent on a three-item lunch order. Then they figure out how much change this student would receive if the order were paid for with $5.00. They show their strategies for each problem on Student Sheet 8, One Lunch Order. Students might use a copy of Coins and Bills to help them solve this problem.

 **Extension**

Shopping at Home   Encourage students to estimate the cost of items in the grocery basket or cart when they are shopping with their families. Students may report back their experiences about how close their estimates matched the total bills.

Making Change

What Happens

Students develop procedures for making change. They learn to count up as a way of making change. They do a number of transactions that involve making change. Their work focuses on:

- calculating change
- counting up to make change

Materials

- Play coins and bills for each student
- Coins for the overhead projector (optional)
- Student Sheet 9 (1 per student, homework)
- Coins and Bills (1 per student, homework, optional)
- Overhead projector (optional)

Activity

Change from $1.00

Distribute play money to students.

I went to the store and bought two things that totaled one dollar. The first thing, a box of raisins, cost 57¢. How much did the other thing cost?

Students share answers and strategies for how they solved this problem. Next, ask students how you could record what happened in a number sentence (for example, .57 + .43 = 1.00, or $1.00 − $.57 = $.43). Record their ideas on the board or overhead.

Here is a little different problem: I went to the store and bought a box of raisins that cost 57¢. I gave the clerk $1.00. How much change did I get back?

It might be obvious to some students that this problem is quite similar to the previous one; other students may not see the connection. Students may use play money, calculate the change mentally, or use calculators to figure it out. Strategies might include subtraction, counting up from 57¢ to $1.00, or adding coins to 57¢ to make $1.00.

Counting Up to Make Change Ask students to think about times when they have paid for something in a store or a restaurant and have needed change back. Can they remember how the cashier figured out their change? Many cash registers automatically calculate change, and students might be familiar with this. Other students might also have noticed that some cashiers also "count back" the change due. If a student does not bring this up, use an example to explain how this is done.

The other day I bought raisins for 57 cents and gave the clerk $1.00. When the clerk gave me my change, she said "57¢" and then gave me 3 pennies and said "58, 59, 60." Then she gave me a dime and said "70," gave me a nickel and said "75," and gave me a quarter and said "$1.00." [*Demonstrate by using money on the overhead or by giving change to a student.*] **Can you explain what she was doing?**

After students share their thoughts, have them calculate the amount of change you received.

Altogether I received 3 pennies, a dime, a nickel, and a quarter. How much change did I get back?

Have students work in pairs to calculate the amount of change back for the following problem:

I bought something that cost 34¢ and gave the clerk $1.00. How much change should I get back?

Emphasize that the idea of counting up is to get to multiples of 5, 10, and 25 cents—amounts for which it's easy to finish making a dollar.

Ask for a pair of volunteers to demonstrate how they figured change back for this problem. One can be the clerk who counts back the change, and the other can be the customer who verifies the amount of change received. One difficulty that some fourth graders encounter with this method of counting up is that though they can give change accurately by counting up using the correct coins, they may not understand that they then have to count the coins to figure out the amount of change received. See the **Teacher Note**, Keeping Track of Addition and Subtraction (p. 26).

Activity

Teacher Checkpoint
Spaghetti and Sauce

Present one more problem to students:

Suppose you bought a package of spaghetti and a jar of sauce for $3.62 and gave the clerk $5.00. How much change should you get back? Write down how you solve this problem. Include a number sentence that shows what you did, then write a statement about how you calculated the change.

❖ **Tip for the Linguistically Diverse Classroom** Instead of having them write a statement, have limited English proficient students use coins to show you how they solved the problem, draw pictures of coins, or use Coins and Bills (p. 79) to cut out and paste on their sheets.

As students are working, circulate around the room and observe how they are calculating the change. Ask them how they determined the amount of change and what coins they used. Some students may choose to use the strategy of counting up, while others might prefer a different method. Don't insist the students use the counting-up procedure you just demonstrated but acknowledge all procedures that make sense. When most students are finished, collect their papers so you can refer to them later.

Looking at Student Work This set of papers can give you an idea about how your students are comparing two quantities and what strategies they are using to calculate the difference between them. If you feel your students could benefit from more problems of this type, you may want to do more examples together.

If there is time at the end of this session, students can begin working on Student Sheet 9, How Much Change?, which they will finish for homework.

Session 6 Follow-Up

How Much Change? Distribute Student Sheet 9, How Much Change?, to each student. Some students may also want to take a copy of Coins and Bills to cut out and use if they do not have access to change at home. They may use any strategy they wish to find the change.

 **Homework**

Exact Change Ask students to explain why some people give the clerk $1.02 instead of $1.00 when an item costs 52¢. What other examples can students think of where it would make sense to add change to the bills used to pay for an item.

 **Extension**

Keeping Track of Addition and Subtraction

During mental computation, especially when several numbers are involved, students should get into the habit of jotting down intermediate steps so they don't lose track of their procedures. For example, suppose you are adding the four numbers below. You might jot down the partial sums like this as you add from left to right:

1540	3000 (the sum of the 1000 and 2000)
347	1700 (the sum of 500, 300, and 900)
2063	150 (the sum of 40, 40, 60, and 10)
+ 918	18 (the sum of 7, 3, and 8)

Now you can easily add the partial sums in your head and write down the result.

Encourage students to jot down their intermediate steps. Show them what you mean by recording students' intermediate steps as they tell you their strategies. Encourage students to jot down their steps in order to remember their approaches and explain them. For example, Karen is adding up several prices. Here is what she might think and write:

$2.07 + $1.49 + $1.99

Karen thinks:	*She records:*
I'll make $1.49 into $1.50 and $1.99 into $2.00.	$3.50
Then I'll add 2 more dollars from the $2.07.	$5.50
I'll add on the 7 cents from the $2.07.	$5.57
Now I'll subtract the 2 extra cents I put on at the beginning.	$5.55

Keeping track of these intermediate steps helps Karen organize her thinking and provides a record so she can later explain her reasoning.

For problems of comparison, a number line notation will help some students keep track. For example, suppose the class is comparing right and left handfuls of beans: "I have 84 beans in my right hand and 142 in my left. How many more could I hold in my left hand?"

A student could write:

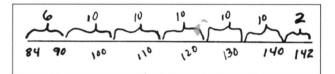

Subtraction strategies make sense for comparison, too. Here is what one student did to document how she found out how far it is from 638 to 1000:

1000 – 600 = 400
400 – 38 is like 400 – 40 = 360
add back 2, since it's really 38
360 + 2 = 362

Students need to know that "taking notes" while they are working on their own procedures is acceptable, valued, and necessary. Otherwise, students often hide their jottings or try to keep everything in their heads because they think they should write down only answers or certain accepted procedures. When you do problems with the whole class, demonstrate different ways of recording what you are doing and solicit other ideas from students.

Shopping Smart

What Happens

Students finish working on their choice activities. They share their approaches for working on the first grade book order. They do an assessment task that involves estimating amounts, finding an exact total, and making change. Their work focuses on:

- estimating and calculating total amounts
- making change
- writing number sentences to represent problems

 Ten-Minute Math: Likely or Unlikely? Once or twice in the next few days, do Likely or Unlikely? with your students. This activity can be done during any free ten minutes of the day. This time include two additional categories or headings: Very Likely and Very Unlikely.

Ask students to write some statements that are very likely or very unlikely. Discuss how a statement that is very unlikely differs from one that is just unlikely.

Students consider the statements they have written using these new categories—deciding which ones are likely, unlikely, very likely, or very unlikely.

How can you change a likely statement into a very likely statement?

Keep the list posted in the classroom so it can be added to each time you do this activity.

For complete directions and variations, see p. 58.

Materials

- Choice Time materials
- Transparencies of several students' first grade book orders
- Student Sheet 10 (1 per student)
- Calculators (1 per student)
- Play coins and bills
- Overhead projector
- Student Sheet 11 (1 per student, homework)

Activity

Completing Choice Time Activities

Students spend the majority of time in these two sessions completing choice activities: Buying Groceries, First Grade Book Order, Close to 100, and Beat the Calculator. Plan to have a class discussion (about 15 minutes) at the end of Session 7 and plan to do an assessment task sometime during session 8.

Tell students that tomorrow will be the final day of this investigation and the final day of working on the Choice Time activities. Acknowledge that some students may not be able to do all the activities but that it is more important they work in depth on several of the choices.

Explain to students that at the end of today's session, the whole class will discuss the First Grade Book Order and that they should be sure to complete this activity in preparation for the discussion.

Class Discussion: First Grade Book Orders

Begin this discussion by having students share the strategies they used for choosing the items in their book orders:

Without telling which books you ordered, how did you go about choosing the items for your order? How did you determine whether you were close to $100.00? How did you calculate the total amount?

Have students share the reasons for their decisions and their strategies for estimating and calculating total amounts.

On the overhead projector, place a transparency of a student's book order, with the total covered up. Ask students to estimate the total and to share their estimates and strategies for estimating with the students sitting near them. When groups are ready, show them the actual calculated total.

Repeat this activity with two or three other students' book orders.

Assessment
Camping Supplies

Plan twenty minutes in Session 8 to bring the whole class together to work on the assessment task. Distribute a copy of Student Sheet 10, Camping Supplies, to each student. Coins, bills, and calculators should be available. Tell students that you would like to get an idea about how they are thinking about estimating totals, using the calculator to figure totals, and determining how much change is due.

Explain that you are interested in knowing how they solved each of these problems and that part of their task will be to explain this using words and number sentences. They can use calculators if they like, but they should describe what they entered on the calculators as part of their explanations. They should not, however, use the calculators to do the first part of the problem, which requires them to make an estimation of the total.

❖ **Tip for the Linguistically Diverse Classroom** Have limited English proficient students respond to this assessment orally. Read each part aloud, then allow them time to solve the problem. For numbers 2 and 3, observe how students are using materials and what kinds of computations they record in order to arrive at final answers.

Looking at Students' Work As you look through students' papers, think about the following questions:

- Does the student use an effective strategy for making a reasonable estimate?
- Does the student record the problem and its solution using addition and subtraction notation?
- What strategy or strategies does the student use to solve these problems?
- Does the student use materials, such as coins and calculators, appropriately to find the solution?
- Can the student communicate clearly about the way he or she solved the problem?

Sessions 7 and 8 Follow-Up

Art Supplies For homework, students solve problems about a student's shopping trip for art supplies. Students record their work on Student Sheet 11, Art Supplies, so that someone else could understand the strategies they used to solve these problems. Students might use the cutouts of Coins and Bills already at home.

 **Homework**

How Far? Measuring in Miles and Tenths

What Happens

Sessions 1 and 2: Miles and Tenths of a Mile
Students consider distances in miles and tenths of miles. They find out how many feet are in $1/10$ of a mile and think about how long that distance is by using more familiar lengths. Using distances that people have run, they figure out how to combine tenths of miles and how to determine weekly mileage for a runner. They make their own fictional running log, with a total of at least 10.5 miles spread across a week.

Session 3 (Excursion): How Far Is $1/10$ of a Mile?
Students work with partners or in small groups to measure out $1/10$ of a mile in the school yard or somewhere else in the school or its vicinity. They might use measuring tools, such as lengths of string or adding machine tape, or they might pace out the distance.

Session 4: A Tour of Our Town   Using a map of their city or town, students plan a tour of the area. They calculate the distances between familiar landmarks using tenths of a mile. Each writes a description of his or her tour.

Mathematical Emphasis

- Estimating local distances in miles and tenths of miles; developing a sense of about how long a mile and $1/10$ of a mile are

- Comparing and combining decimal numbers and finding differences between these numbers

- Seeing the relationships of decimal parts to the whole

- Measuring distances on maps using a scale

- Becoming familiar with common decimal and fraction equivalents (for example, that a quarter of a mile is written as .25 mile)

- Estimating and calculating sums of quantities that include decimal portions

What to Plan Ahead of Time

Materials

- Calculators: 1 per student (Sessions 1–4)
- String: 3–4 rolls (Session 3)
- Adding machine tape: 3–4 rolls (Session 3)
- Yardsticks: 3–4 (Session 3)
- Scissors (Session 3)
- Map of your city or town with a scale of miles: 1 per group (Session 4)
- Index cards: 1 per student (Session 4)
- Rulers: 1 per pair (Session 4)

Other Preparation

- Duplicate student sheets and teaching resources, located at the end of this unit, in the following quantities. If you have Student Activity Booklets, no copying is needed.

For Sessions 1–2

Student Sheet 12, Runners' Logs (pages 1 and 2) (pp. 80–81): 1 per student

Student Sheet 13, Making a Running Log (p. 82): 1 per student

Student Sheet 14, Calculator Skip Counting (p. 83): 1 per student

For Session 3

Student Sheet 15, How Far Is $1/10$ of a Mile? (p. 84): 1 per student

Student Sheet 16, My Measuring Method (p. 85): 1 per student (homework)

For Session 4

Student Sheet 17, A Tour of Our Town (p. 86): 1 per student

Student Sheet 18, My Tour (p. 87): 1 per student (homework)

- Clock the distance from your school to a familiar place, not more than 5 miles from school. If you do not have a car, see if a friend or a student's parent can do it. This should be clocked on an odometer, with the actual odometer reading recorded.

- Obtain a map of your city or town. It should have a scale of miles on it. If you live in a large city, you might choose a map of the neighborhood surrounding the school. The city hall or real estate agencies are potential resources for local maps. If your map is small enough, make several copies (1 per group of students is ideal).

Miles and Tenths of a Mile

Materials

- Student Sheets 12–14 (1 of each per student)
- Calculators

What Happens

Students consider distances in miles and tenths of miles. They find out how many feet are in ¹/10 of a mile and think about how long that distance is by using more familiar lengths. Using distances that people have run, they figure out how to combine tenths of miles and how to determine weekly mileage for a runner. They make their own fictional running log, with a total of at least 10.5 miles spread across a week. Their work focuses on:

- measuring using miles, tenths of miles, and quarters of miles
- becoming familiar with common fraction/decimal equivalents
- visualizing distances such as ¹/10 of a mile
- adding numbers that include whole numbers, tenths, and quarters

Activity

Introducing Parts of a Mile

Report to students that you, a friend, or a parent recorded the mileage from school to a familiar landmark. You might say:

Yesterday when I left school, I drove to ____ and found out how far away it is from here. Do you have any guesses as to the distance?

Students will at first probably guess in terms of whole miles, though some may say such things as "between 2 and 3 miles" or "2 ½ miles."

To figure out how far it was, I wrote down the readings on my odometer. When I left school, the odometer looked like this: ___. When I reached ____, it looked like this: ___. [*Write the two odometer readings where students can see them.*]

How would you read these two numbers? What whole numbers are these numbers close to?

Some students will be unfamiliar with reading numbers like 25432.1 because there are no commas. They may want to put the commas in to show where the thousands are.

Other students may have difficulty reading this number because it includes decimals. Many students will read the decimal as "point one." Accept this, but also introduce the vocabulary of tenths of a mile. Connect this with students' work with decimals and money in Investigation 1. Just as decimals in money amounts represent part of (or less than) one dollar, a decimal in this context means part of (or less than) one mile.

Once they have successfully determined what the numbers are, ask them to estimate how far you actually went.

About how far did I go? How did you figure that out?

Encourage students to estimate the total distance—for example, "Well, you started at 25432.1, which is really close to 25432. Then you went to 25435.4, which is almost half a mile more than 35, so you went around 3 ½ miles, but a little less."

Running Distances

Ask students if they have ever run in a race or a fun run. For those who have, ask what distances they ran and write these on the board. Discuss how long it took to run this far or have students describe the race course so others might get a sense of how far the distance really was. (If no one has run, you might want to tell students that common race distances for children their age are .5 mile, 1 mile, and 3.1 miles [5 kilometers]. If you've run in a race, you might want to write down your race distance and describe your experience.) See the **Teacher Note**, Decimal Events (p. 35), for ideas about community events in which students may participate.

Usually runners try to prepare for a race by running a certain number of miles each week. In a week, they try to run three times as much as the distance of the race. If they were running a 3.1-mile race, how many miles would they try to run in a week?

After the class has given some suggestions for figuring this out, distribute Student Sheet 12, Runners' Logs, and ask students to work with partners on the problems. On the sheet are two "logbooks" that show how far different runners ran each day and how they felt. Your students' task is to determine how far these runners went in a week and if they ran far enough (at least 9.3 miles) to prepare for the upcoming race.

Observing the Students   As students are working on these problems, talk with them about their ways of adding decimals. Encourage students to develop their own workable way of combining the numbers and then to use a second method to check their results. Strategies students use may include adding from left to right—first the ones, then the decimal pieces; using a calculator; using nearby landmarks, adding, and then adjusting the total accordingly (see the **Teacher Note**: Three Powerful Addition Strategies, p. 10). For example, when adding 1.6 + 2.7, some students might combine 1 + 2 = 3, then combine .6 + .7 = 1.3, and finally add 3 + 1.3 = 4.3.

When students have finished, discuss with them how they solved the problems. During the discussion, pay particular attention to the following:

- Are students able to add the decimal pieces accurately? Do they see that numbers like .6 and .7, when added, give a total of 1.3 (rather than .13)?

- Do students know how to combine numbers like .25 and .1? If not, remind them that this process is like adding a quarter and a dime. You may want to ask them some questions:

  If I ran 2.2 miles and you ran 2.25 miles, who ran farther? Why do you think so?

  Which is farther—2.5 miles or 2.25 miles? Why do you think so?

- Do students get reasonable results? All their sums should be in the right ballpark. If they are getting sums such as 94.2, they should go back to the individual numbers of miles and look at the whole number of miles before adding.

Activity

Teacher Checkpoint

Making Your Own Log

Each student will make his or her own fictional running log.

We are each going to make our own imaginary running logs. Imagine you are going to run a 3.5-mile race. [*Write this number on the board.*] **About how far would you have to run each week to get ready for this race? How would you do it? Would you run the same amount each day or run different amounts on different days?**

Discuss different approaches to breaking up the mileage. You might want to establish a rule that you can't run all the mileage in one day. The class may want to decide that you can have only one or two rest days in a week. Once you've discussed the ground rules, distribute Student Sheet 13, Making a Running Log. Students work individually to make up their own fictional running logs, making sure they have at least 10.5 miles in their weeks.

❖ **Tip for the Linguistically Diverse Classroom** Offer limited English proficient students the option of entering their comments in their native languages or with illustrations. For example:

Day	Mileage	Comments
Wednesday	1.25 miles	

Collect these papers and review them to determine how your students are doing at combining numbers that involve decimals.

Looking at Students' Work

- What strategies are students using to combine decimals? Are they able to add decimals accurately?

- How do students combine numbers like 1.75 and .5? Do students consider whether their results are reasonable? Encourage students who are having difficulty to focus on the whole number of miles and to estimate the amount before adding.

- How do students keep track of their work? How do they represent their work? Do they use addition notation?

Calculator Problems

If students finish with their running logs, have them work individually with calculators and Student Sheet 14, Calculator Skip Counting. They try to find what numbers less than 1 they can skip count by to get 1, 2, or 3. Students can skip count by adding (.2 + .2 + .2 + . . .) or by using the constant key (0 + .2 = = = = . . .)

Decimal Events

Be on the lookout for the many special events— bike rides, walks, and runs in your neighborhood—that might be of interest to students and families. Many charities sponsor special "walks" where participants collect pledges ahead of time. The pledges are often based on how many fractions of a mile a participant completes. These are perfect opportunities for students to incorporate math and community service. If you anticipate such an event in your community, you may want to encourage students to participate. If students mention they have done such a walk, ask them to talk about how far 5.5 miles (for example) felt, how long it took to walk the distance, whether there were any markers on the course to show how far they had gone, and so on. Of course, figuring out how much pledge money they earned based on the number of miles completed is an opportunity for students to do repeated addition or multiplication with decimals. Both fractions of a mile and fractions of a dollar come into play.

In addition to charity events, most communities have occasional fun runs and bike rides for students and their families. (Distances of 3.1 and 6.2 miles are very common for these events,

because they correspond to the metric measurements of 5 and 10 kilometers. This may be a good opportunity to discuss how the metric system is used in most countries in the world and to compare the metric system of measuring distances with our own system.) Again, these are opportunities for students to get first-hand experience with various distances. How long does it take to ride 3.1 miles? To run or walk this distance? Actually participating in an event like this makes the distances more concrete for students. Encourage students to get involved and to keep logs of their training miles, similar to the ones they completed for the activity in Session 2. Keeping track of their mileage over time is an excellent way to get practice with adding decimals. If students run some of their practice sessions on a track, the distances are typically measured in quarter miles. Other distances they run will be measured in tenths of a mile. They will quite naturally come across the problem of how to add distances such as .25 and .8. Most students who have run these distances understand—on a physical as well as an intellectual level—that the total is not .105 or .15 mile.

How Far Is ¹⁄₁₀ of a Mile?

Materials

- String (3–4 rolls)
- Adding machine tape (3–4 rolls)
- Yardsticks (3–4)
- Scissors
- Calculators
- Student Sheet 15 (1 per student)
- Student Sheet 16 (1 per student, homework)

We hope you will find a way for students to undertake this activity since it is important that they get some sense of the length of these distances and be able to visualize them, rather than just manipulate them on paper. Before beginning this activity, choose the place or places where you will do your measuring. Depending on your situation, you might want to measure how many laps of the gym are in ¹⁄₁₀ of a mile, whether the perimeter of the building is close to ¹⁄₁₀ of a mile, or how much of the playground is ¹⁄₁₀ of a mile. Different teams can work in different places, or every team can do the same measurement.

What Happens

Students work with partners or in small groups to measure out ¹⁄₁₀ of a mile in the school yard or somewhere else in the school or its vicinity. They might use measuring tools, such as lengths of string or adding machine tape, or they might pace out the distance. Their work focuses on:

- determining and visualizing how long ¹⁄₁₀ of a mile is
- inventing ways to measure large distances

Activity

How Far Is ¹⁄₁₀ of a Mile?

We talked a lot about tenths of miles in the last couple of days. Now we're going to figure out how far it really is. Many distances are measured in tenths of a mile.

Write 5280 on the board and explain to students this is how many feet there are in one mile. As a class, students discuss how they would figure out ¹⁄₁₀ of a mile.

When students have figured out that ¹⁄₁₀ of a mile is 528 feet, explain that they'll be measuring this distance.

If we were going to start at the front door of the school and go to the end of the playground, would it be ¹⁄₁₀ of a mile? How much farther do you think we would have to go to get to ¹⁄₁₀ of a mile?

Distribute Student Sheet 15, one per student.

❖ **Tip for the Linguistically Diverse Classroom** Pair limited English proficient students with English-proficient students to complete this student sheet.

In pairs, students spend a few minutes planning how they will measure $\frac{1}{10}$ of a mile using the materials available in the classroom.

Make available any tools you have that would give students a way to measure something longer than a yardstick, such as string or adding machine tape.

Strategies that students might use include these:

1. They might measure out a piece of string or adding machine tape that is some distance longer than a ruler or yardstick (perhaps 10 feet, 25 feet, or even 52.8 feet). Then they use this string or tape as a measuring tool. Students will have to figure out how many times they need to put down the tape or string in order to get to $\frac{1}{10}$ of a mile.

2. They might choose some distance to measure off in the hallway or outside (10 feet, 25 feet, or the like) and figure out how many paces it takes to walk that distance. How many of these distances would they need to get to $\frac{1}{10}$ of a mile? They pace off this many of their chosen distance.

3. Some students may have the patience to use a yardstick to measure $\frac{1}{10}$ of a mile. They will have to figure out how many yards are in $\frac{1}{10}$ of a mile.

Students might invent a variety of techniques other than those above, and it is important that they develop their own methods.

Have a brief conference with each pair of students to make sure they have their plans worked out and materials ready. Make sure students have made plans to keep records and mark off distances as they go along. For example, if students use method 1 above, urge them to make a mark of some kind on the ground or make a tally each time they lay out their string or tape. There is nothing more frustrating than getting toward the end of this task and realizing they've lost track of the count!

Measuring $\frac{1}{10}$ of a Mile and Discussing the Findings Students can figure out their methods in class, with their partners, then do the actual measuring during recess, after school, or during a class walk. What you do with your class will vary depending on the location of your school, whether you have easy access to outside areas, and so forth. Students may need to do the work on their own time in order to avoid collisions between teams in the chosen location.

When the class meets again, discuss the results. Was $\frac{1}{10}$ of a mile longer or shorter than they imagined? Where did they get to? How many laps of the schoolyard or gym equaled $\frac{1}{10}$ of a mile? If students did this again, how would they do it differently? See the **Dialogue Box**, Measuring $\frac{1}{10}$ of a Mile (p. 39), for examples of ways students measured $\frac{1}{10}$ of a mile.

Session 3 Follow-Up

 **Homework**

My Measuring Method   On Student Sheet 16, My Measuring Method, have students write brief reports on what measuring methods they used, how they worked, and what they found out. Suggest they frame their reports around these questions:

> What was your plan?
>
> What did you do?
>
> What did you find out?

They should also include sketch maps showing their results.

Measuring 1/10 of a Mile

This dialogue took place as students engaged in the activity How Far Is 1/10 of a Mile? (p. 36).

Now that you've had a chance to measure out 1/10 of a mile, I'd like a few of you to tell us how far it was. When you tell us about it, focus on these things [*writing on board*]: **What was your plan? What did you do? What worked or didn't work?**

Luisa: We measured the counter with the science stuff on it. It was 17 feet long.

Rebecca: So we made a tape 17 feet long. Then we thought, "How many of these would go in 1/10 of a mile?" We needed 31 tapes.

How did you figure that out?

Rebecca: We used a calculator. We said, "How many 17's do you need to make 528?" That was a little hard, but we just tried different numbers, timesing them by 17 until we got close.

Marci: Well, 31 tapes actually makes 527 feet, but it's close enough.

Luisa: Then we decided to see how many tapes it would take to go across the field the long way. We put it down 15 times to go across the field. We almost lost track, so we had to start over to check, but . . .

Joey: Didn't your tape break?

Luisa: No, yours probably broke because it was too long.

Rebecca: But then we figured when we got to the end of the field—well, it was 15, and two 15's make 30, so all we had to do was go back again, and we'd just about have 1/10 of a mile!

Marci: We still needed one more tape, but really it's half of that because we went both ways. So we say 1/10 of a mile is down the field and back and a little bit more.

Alex, how did your group do it?

Alex: We decided to measure around the base-ball diamond. We started at the backstop and just kept putting it down . . .

What were you putting down?

Shiro: We used the meter stick. We started with a tape, but it was too long and ripped when the wind got it.

Emilio: Yeah, so we figured out that 162 meter sticks made 528 feet.

How did you know that?

Emilio: Well, there's 3.25 feet in a meter, so we said how many 3.25 feet in 528? We knew we had to divide 528 by 3.25, but that's pretty hard, so we did it on the calculator.

Shiro: The hard part was keeping track, because 162 is a lot. We made a slash each time we put down the stick.

Teresa: How far did you get?

Alex: Around the field one whole time, then to the stop sign. It's actually a little bit farther than that.

Shiro: Yeah, it's a meter farther. We think. But we're not absolutely sure because we might have lost track.

Tyrone: Can we go next? We did it a different way.

Sure.

Tyrone: Well, it was really fun. We used string because we didn't want it to break. We used a big piece of string, 50 feet long.

Vanessa: Yeah, because it wouldn't take so many. We already knew that 50 feet would be about 10 times and a little more to make 1/10 of a mile.

Jesse: Actually it's 10 1/2 of our string to make 1/10 of a mile.

Shoshana: So we just went out the east door and decided to keep going straight until we saw how far 1/10 of a mile was.

Continued on next page

Vanessa: But it wasn't that easy! We started going from the door out to the stop sign. That was 4 strings. But then we had problems. We ran into some bushes.

Jesse: We didn't know what to do when you couldn't go straight.

So what did you do?

Shoshana: Then we decided to go in a different direction. That worked for a while, but then we had to go around a car that was parked in the parking lot.

Tyrone: [*showing his sketch*] So we'd say 1/10 of a mile is from the east door, out to the stop sign, then around to the parking lot, and most of the way across the parking lot. Except we don't know exactly because we had to zigzag a little for that car.

A Tour of Our Town

What Happens

Using a map of their city or town, students plan a tour of the area. They calculate the distances between familiar landmarks using tenths of a mile. Each writes a description of his or her tour. Their work focuses on:

- combining and comparing decimal amounts
- determining distance using a scale of miles

Materials

- Maps of your city or town with a scale of miles (1 per group)
- Rulers (1 per pair)
- Index cards (1 per student)
- Student Sheet 17 (1 per student)
- Student Sheet 18 (1 per student, homework)

Activity

City Landmarks

Introduce this session by asking students to think about some of the important and/or interesting landmarks in your city or neighborhood.

Suppose you had some friends who had just moved here, and they asked you to give them a tour of the city (neighborhood) so they would know where things were located. What are some of the places you think they might like to visit or should know about?

Note: Depending on where you live and what type of map you were able to obtain, you will need to adapt this session to fit the needs of your situation. If you live in a large city, you might choose to have students focus on the neighborhood around the school. If you live in a smaller city or town, it might make more sense for students to plan a tour of the entire town.

Make a list of students' suggestions. Important locations might include the school, a favorite playground, favorite gathering places (like the mall or movie theater), historical landmarks or museums, the post office, and the library.

❖ **Tip for the Linguistically Diverse Classroom** Include rebus pictures for the places named on the lists. For example:

An addressed letter next to the words *Post Office*

Books next to the word *Library*

Using the Map Distribute maps of the area to students. Group students according to how many copies you were able to obtain. If necessary, a group of up to five or six students can share a map.

Give students some time to become familiar with this map. Suggest they try to locate a few familiar places, starting with their school. When they have found it, one person in the group should mark it on the map and label it.

The amount of orienting you will have to do with the map will depend on how much experience students have had with looking at maps.

Determining the Scale of Miles Ask students if anyone is familiar with a scale of miles, what it represents, and what it is used for. If no one knows, offer this explanation.

One piece of information people often want to know when they are using a map is how far away places are from one another. On a map, the scale of miles will tell you how many miles from one place to another. Many times mileage scales are figured in inches or centimeters. For example, 1 inch on the map might equal 10 miles. Usually, when maps show large areas such as the United States or the world, the scale of miles is large, so 1 inch might equal 150 miles. When a map represents a smaller area, such as a neighborhood or a city, the scale of miles is much smaller. One inch might represent 1 mile or even ½ mile.

Students locate the scale of miles on their maps and determine the scale. Write the scale information on the board. For example: 1 inch = .6 mile. Many city or neighborhood maps will be calculated in tenths of a mile. Relate this information to the previous sessions where students were measuring $\frac{1}{10}$ of a mile and planning runs using $\frac{1}{10}$ of a mile. If the scale on your map is in whole miles (for example, 1 inch = 1 mile), students will most likely need to use fractions of a mile (for example, school to post office is 3$\frac{1}{2}$ inches; that's 3$\frac{1}{2}$ or 3.5 miles). Students can record fractions of a mile with either fractions or decimals.

Sometimes when people are using the scale of miles to figure out distance on a map, they use their fingers to estimate the size of the scale and then count out how many times that amount fits between the two points. Other times people might use a ruler to measure the distance. You can also use a strip of paper that matches the size of the scale.

Pass out index cards and have students mark the scale of miles along one edge. They can use this to measure the distance by figuring out how many times that length fits between the two points.

Choose a familiar location near your school. Have students locate it on their maps and label it. Using either a ruler or their scale strip, students figure out the distance between these two places.

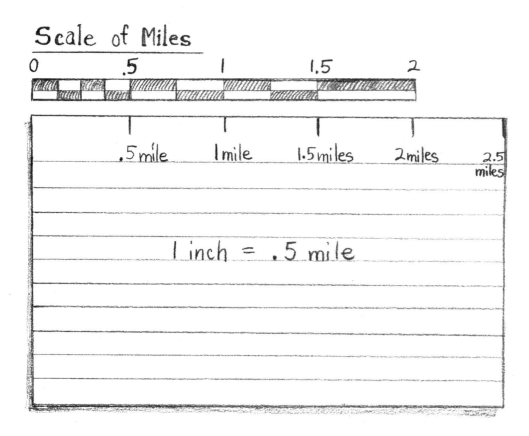

Record the distances on the board and have groups of students explain how they determined the distance. See the **Dialogue Box**, How Far Is It to the Post Office? (p. 45), for examples of how students might calculate the distance using scale.

Activity

Planning the Tour

You and your partner(s) will be planning a tour of our city (neighborhood). In addition to choosing places to visit and locating them on your maps, you will need to figure out the distances between the places on your tour. You should record all this information on Student Sheet 17, A Tour of Our Town. When you are finished, calculate the total number of miles that your tour covers.

Students work for the remainder of this session on planning their tours. As students finish, they can begin the homework assignment. Post the maps and descriptions of students' tours around the classroom.

Session 4 Follow-Up

 **Homework**

My Tour   Students list places on their tours and write descriptions of the tours they have planned on Student Sheet 18, My Tour. Suggest they write from the perspective of a tour guide. They should describe each place the tour visits and what might be interesting or special about that place.

How Far Is It to the Post Office?

Using a map, students determine the distance between their school and the post office in the activity City Landmarks (p. 41).

Nadim: First I measured from the school to the post office and that was 3 ½ inches. The scale is .6 miles for every inch, so I wrote .6 + .6 + .6, and I figured that out by adding. I got 18, then I had to add half of that for the half inch, so that was .3. I think it's 18.3 miles from school to the post office.

Karen: I did the same thing: I added .6 + .6 + .6 + .3, but I got only 2.1 miles.

Nadim: Hmm. The post office isn't very far from the school, it's near my house, so I think 2.1 miles sounds better than 18.3.

So you think Karen's answer is more reasonable than yours?

Nadim: Yes, but we added the same numbers.

Rebecca: I see what happened. You added the three 6's, but you forgot the decimal point, so you ended up with 18 instead of 1.8. That's what made your answer way too big.

Nadim: Okay, so if I change it to 1.8 and then add the .3, it will come out to 2.1. That sounds better.

Anyone have any other ways of figuring the distance?

David: Well, I used the index card like a ruler. I had the scale on the card, and each of those sections was .6 miles, so I lined one end up at the school and then just made a mark on the map where the scale ended. Then I repeated that until I got to the post office. It took about 3½ scale lengths.

And what did you do then?

David: I used the calculator and added .6 + .6 + .6. That was 1.8. Then I wasn't sure what to do next.

Karen: So I told him he had to add in the half, and half of .6 is .3. It's just like half of 6, but it's in decimals.

David: I'm still not sure about that part of it, but I know that it's around 2 miles to the post office.

Using a scale to measure short distances can be tricky, especially when the scale represents part of a mile instead of a whole mile. You have some good strategies for figuring out this distance.

Calculating Longer Distances

What Happens

Session 1: Close to 1000 As an introduction to combining and comparing large numbers, students play Close to 1000, a version of the Close to 100 game they played in Investigation 1. The session ends with a discussion of strategies used to combine and compare larger numbers.

Sessions 2, 3, and 4: A Trip Around the United States In these three sessions, students plan a trip around the United States. Using the mileage scale on their maps they determine distances between the cities they want to visit. They plan a trip to between four and ten places, but they cannot travel more than 10,000 miles. They keep travel logs of their trips.

Mathematical Emphasis

■ Measuring distances on maps using a scale

■ Comparing and combining numbers in the hundreds and thousands

■ Using standard addition and subtraction notation to record combining and comparing problems

What to Plan Ahead of Time

Materials

- Calculators: 1 per student (Sessions 2–4)
- Tape measures: 1 per pair
- Wall map of the United States (Sessions 2–4)
- Strips of paper the length of the scale on the map (optional) (Sessions 2–4)
- Inch cubes if the scale on the map is in inches (optional) (Sessions 2–4)
- Travel guides or brochures for the United States: 2–3 per pair (Sessions 2–4)
- Overhead projector (optional)

Other Preparation

- Duplicate student sheets and teaching resources, located at the end of this unit, in the following amounts. If you have Student Activity Booklets, no copying is needed.

Session 1

Student Sheet 19, Close to 1000 Score Sheet (p. 88): 1 per student (class), 1 per student (homework)

Numeral Cards (pp. 92–94): 1 deck per pair (class), 1 deck per student (homework)

How to Play Close to 1000 (p. 91): 1 per student

Sessions 2–4

Student Sheet 19, Close to 1000 Score Sheet (p. 88): 1 per student (homework)

Student Sheet 20, Travel Log: U.S. Trip (p. 89): 4–10 per student, depending on how many places they visit

Student Sheet 21, U.S. Map (p. 90): 1 per student

How to Play Close to 1000 (p. 91): 1 per student (homework)

- Make five or six extra decks of Numeral Cards (pp. 92–94) for Close to 1000 for classroom use.

- Prior to beginning this investigation, collect travel brochures and guides to places around the United States. These are usually available from a travel agency. Many libraries also have a selection of tour guide-books. Students can check at home for travel books.

Close to 1000

Materials

- Numeral Cards (1 deck per group of 2–3 students, class; 1 deck per student, homework)
- Student Sheet 19 (1 per student, class; 1 per student, homework)
- How to Play Close to 1000 (1 per student)
- Overhead projector (optional)

What Happens

As an introduction to combining and comparing larger numbers, students play Close to 1000, a version of the Close to 100 game they played in Investigation 1. The session ends with a discussion of strategies used to combine and compare larger numbers. Their work focuses on:

- finding combinations of numbers that are close to 1000

 Ten-Minute Math: Likely or Unlikely? Once or twice in the next few days, do Likely or Unlikely? with your students. This activity can be done during any free ten minutes during the day. This time use the categories, or headings, More Likely and Less Likely.

Introduce the element of comparison with statements using *more likely* or *less likely*. For example:

It is more likely that it will rain tomorrow than that it will snow.

It is less likely that I will see a mouse today than a dog.

Activity

Playing Close to 1000

(If you are doing the full *Investigations* curriculum, your students will have played Close to 1000 in the unit *Landmarks in the Thousands*. If you feel your students are comfortable with this game, it may be appropriate to begin with Session 2.)

Introduce this game as a new version of Close to 100, a game students will be familiar with from Investigation 1. The one difference from Close to 100 is that players are dealt eight cards, and they use six of them to make two three-digit numbers. Explain to students that the object of this game is to make two three-digit numbers whose total is as close to 1000 as possible. The scoring is the same as in the original version of Close to 100. Players compute their scores for each round by figuring out the difference between their totals and 1000. After five rounds, they add up their scores; the player with the lowest score wins. They may also choose to use the alternative scoring, keeping track with positive and negative numbers whether the difference is greater or less than 1000 and adding these numbers for their total scores. Distribute How to Play Close to 1000 (p. 91) and a copy of Student Sheet 19, Close to 1000 Score Sheet, to each student. Students can play in groups of two or three. (**Note:** For future games students can make their own score sheets using Student Sheet 19 as an example.)

Students play Close to 1000 for most of this session. Plan to leave 15 minutes at the end of this session to have a discussion about strategies for combining and comparing large numbers.

As students are playing, observe the strategies they are using to choose the numbers they will combine. Do they make one three-digit number and then figure out how much more they need to get to 1000? Do they focus on the hundreds first as a way of getting close to 1000, then fill in the tens and ones?

Take note of the particular strategies that students are using to combine and compare numbers and integrate these into the discussion at the end of this session.

Activity

Class Discussion: Strategies for Adding and Subtracting Large Numbers

In this discussion, students share their strategies for combining and comparing large numbers. Begin the discussion by presenting this problem to students:

$$518 + 427$$

Suppose these were the two numbers you made in Close to 1000. How can you add them?

Ask for volunteers to demonstrate their strategies on the board or overhead. Solicit as many approaches as you can. Encourage students to record their strategies using mathematical notation (for example, 500 + 400 = 900; 10 + 20 = 30; 8 + 7 = 15; 900 + 30 + 15 = 945) as a way of reinforcing the idea that there are many ways to arrive at a solution.

Imagine you were playing a game called Close to 1500, and you had one number that was 945. What other number would you need to add in order to get 1500?

Give students a few minutes to solve this problem. There is more than one way to solve this. Some students may count up using familiar landmarks; others may count back. Again, suggest that they record their solutions using mathematical notation.

As students discuss both of these problems, listen for evidence that suggests they are thinking flexibly about numbers and have sensible strategies for adding and subtracting that are based on sound number sense and knowledge of important landmarks.

- Are students' estimates reasonable? Can they explain their thinking?
- How do students use landmark numbers to arrive at their solution?
- How do students record and keep track of their work? Do they use standard notation?

Session 1 Follow-Up

 **Homework**

Close to 1000 Students play Close to 1000 with someone at home. They should have one deck of Numeral Cards at home from playing Close to 100. Give each student a copy of Student Sheet 19 (Close to 1000 Score Sheet) and a copy of How to Play Close to 1000.

A Trip Around the United States

What Happens

In these three sessions, students plan a trip around the United States. Using the mileage scale on their maps, they determine distances between the cities they want to visit. They plan a trip to between four and ten places, but they cannot travel more than 10,000 miles. They keep travel logs of their trips. Their work focuses on:

- using a map scale
- measuring distances on a map of the United States
- combining numbers in the hundreds and thousands

Materials

- Calculators (1 per pair)
- Tape measures (1 per pair)
- Wall map of the U.S.
- Paper strips the length of the map scale (optional)
- Inch cubes (optional)
- Travel guides for the U.S (2–3 per pair)
- Student Sheet 19 (1 per student, homework)
- Student Sheet 20 (4–10 per student)
- Student Sheet 21 (1 per student)
- Numeral Cards (1 deck per student, homework)
- How to Play Close to 1000 (1 per student, homework)
- Overhead projector (optional)

Activity

Introduce this activity by explaining that during the next three math classes students will be working with partners to plan a trip around the United States. They will be using a map of the United States to measure the distances between the cities they visit.

Using a Scale of Miles Students will be calculating the mileage from city to city by using the map's mileage scale.

When you are using a map, one of the things you might need to know is the distance between two places on the map. Most maps have a mileage scale you can use to calculate distance. Different maps use different scales. On our map every inch represents ___ miles. [*Substitute the correct information from the map you are using.*] So if I wanted to know how far it was from New York to Chicago, I could measure the distance on the map and then figure out the distance in miles.

Explain to students that for this activity they will be using tape measures to measure the distance between the cities to which they will travel. They will then calculate the number of miles between these cities.

Measuring Distance on a Map

Demonstrate how this is done by choosing two cities on the map and having a student measure the distance with a tape measure. On the board or overhead record information such as:

Scale: 1 inch = 75 miles

New York to Chicago = 9½ inches

(**Note:** This information will vary according to the scale on the map you are using. The approximate distance between New York and Chicago is 713 miles.)

Computing Distance   Remind students of the scale and then ask them to suggest ways they might use this information to calculate the number of miles between New York and Chicago.

Some students might suggest doing repeated addition, adding nine 75's and then figuring out ½ of 75. Some students might use the landmark of 10, figure out 10 x 75 = 750, and then subtract 37½. Other students may need to figure out a one-to-one correspondence and say "One inch is 75, 2 inches is 150, 3 inches is 225, and so on," until they reach 9 inches.

After students have offered some strategies for computing the distance from New York to Chicago, have them work with partners to do the calculation.

As they work, circulate around the room and observe their strategies. If some students are having difficulty understanding what the scale represents, cut out strips of paper that represent the size of the scale (or use inch cubes if your scale is equal to 1 inch) and have the students write the number of miles each strip represents. They can then measure the actual distance on the map by placing these strips end to end.

Activity

A Trip Around the United States

You and your partner will be planning a trip around the United States. You will begin and end your trip in our city. On your trip you may visit between four and ten different places (not counting our city), but you cannot travel more than 10,000 miles. As you plan your trip, you will need to keep a travel log that tells the places you will visit, the mileage you will travel, and a description of what you plan to do in each city. You will also be sketching your trip on a map of the United States.

Distributing Materials   To each student give a copy of Student Sheet 20, Travel Log: U.S. Trip, and Student Sheet 21, U.S. Map. Students will eventually need several more copies of Student Sheet 20, depending on how many places they visit. To each pair also give a tape measure and some travel guides or brochures.

Charting Their Trips   Explain each student sheet to the class and what students are to do. Allow flexibility in the order students perform each task. One way to proceed is for students to record the first leg of the trips on their travel logs (Student Sheet 20) and sketch that distance on their maps (Student Sheet 21). Then they go to the wall map of the United States with their tape measures. Using the scale of miles on the wall map, students find out how far this part of the trip is in inches. At their seats, they figure out how many miles this leg of the trip is and complete the page of their travel logs. Students repeat this process for each leg of their trip, and record their findings on Student Sheet 20.

❖   **Tip for the Linguistically Diverse Classroom**   Pair limited English proficient students with English proficient students and have the partners do Student Sheet 20 together. Students with limited English proficiency can illustrate the comment section of this sheet. For example: (picture of a boy surfing) While we were in California, we went surfing at the beach.

Deciding Where to Go   Before students begin planning their trips, encourage them to spend time with their partners discussing possible places they are interested in visiting. In addition to some of the more well known places, such as Disney World and the Grand Canyon, students might consider visiting special relatives, a place from a favorite story, a place of special interest (Baseball Hall of Fame), or the site of a historical event (a stop on the Underground Railroad). In preparation for their trips, students can look through the travel guides and brochures.

As they plan their trips students will need to keep in mind they must visit at least four places, other than their own city; their trips must begin and end in their own city; and they cannot travel more than 10,000 miles.

Keeping a Travel Log   On the next page is an example of what a page of the travel log might look like after several places have been recorded.

Students should explain how they are figuring out the mileage by using number sentences and words. In this way you can assess the strategies your students are using to combine and compare large numbers.

You may want to use a portion of these sessions to work with small groups who need help in organizing their projects or in developing strategies for comparing and combining large numbers. For these students, you might suggest they begin their trip by visiting places that are close together.

Travel Log: U.S. Trip

From	To	Length	Distance
Boston	Washington, D.C	10 inches	750 miles

How I figured out the distance between these two places:

$75 \times 10 = 750$ I just knew that from multiplication.

Comments:

While we are in Washington, we want to visit the zoo, the White House, the Vietnam Memorial, and the Martin Luther King, Jr., Library. Tyrone's grandmother lives in Washington, so we'll visit her, too.

Total Distance Traveled So Far:

$2548 + 750 =$

$2000 +$
$500 + 700 = 1200 +$
$40 + 50 = 90 +$
8
$= 2000 + 1200 + 98$
$= 3298$ miles

Miles Left:

How far do we have to go to get to 10,000:
$10,000 - 3,000 = 7,000$
$7,000 - 300 = 6,700$
but it's not really 300, it's 298, so add
$6700 + 2 = 6,702$
We have a lot of miles left!

Investigation 3 • Sessions 2–4
Money, Miles, and Large Numbers

Activity

Assessment: Observing Students and Analyzing Their Travel Logs

This assessment task is designed in two parts. The first part consists of observing students as they plan their trips around the United States. The second part of the assessment involves looking at students' travel logs as documents of their work. Each of these pieces of information will give you a fuller sense of their strategies and understanding of addition and subtraction of large numbers.

Observing the Students   During these three sessions, observe each pair of students as they are planning their trip. Consider the following as you observe:

■ Are students working together as a team? How are they deciding where to go? How are they sharing the work?

■ Are students able to measure the distance between two places accurately? Can they compute the mileage based on the scale?

■ Are students using tools, such as the calculator, to solve problems?

Looking at Travel Logs   As you look through each student's travel log, ask yourself the following:

■ What strategies are students using to combine and compare large numbers? Are they accurate?

■ Are students able to represent these relationships using addition and subtraction notation?

■ Can students keep track of their work?

■ Are students able to plan trips of between four and ten places and travel less than 10,000 miles?

■ Can students approximately represent their trips on the map of the United States?

Sharing Travel Logs   Students may be interested in reading one another's travel logs. Choose a place in the classroom where they can be displayed and shared with the class.

Choosing Student Work to Save

As the unit ends, you may want to use one of the following options for creating a record of students' work on this unit:

■ Students look back through their folders or notebooks and write about what they learned in this unit, what they remember most, and what was hard or easy for them. You might have students do this work during their writing time.

■ Students select one or two pieces of their work as their best work, and you also choose one or two pieces of their work to be saved. This work is saved in a portfolio for the year. You might include students' written solutions to the assessment, Camping Supplies (Investigation 1, Sessions 7 and 8), and any other assessment task from this unit. Students can create a separate page with brief comments describing each piece of work.

■ You may want to send a selection of work home for parents to see. Students write a cover letter describing their work in this unit. This work should be returned if you are keeping a year-long portfolio of mathematics work for each student.

Sessions 2, 3, and 4 Follow-Up

🏠 Homework

A Trip Around the United States   Students play Close to 1000 with friends or family members using the alternative method of scoring differences as positive and negative numbers explained at the bottom of the game instructions. Students should have available at home a deck of Numeral Cards, a Close to 1000 Score Sheet, and a copy of How to Play Close to 1000.

⟩⟩ Extension

Famous Journeys   There are a number of extensions that can be done in conjunction with the trip around the United States:

■ Students can plan a trip to a selection of the birth cities of people in their family. They interview parents and other relatives to find out where they were born. Students might collect this information about parents, siblings, grandparents, aunts, or uncles. In some communities, students may not have to go beyond themselves and their immediate families to collect a variety of birthplaces. In other communities, students may have to go back two or three generations to find birth cities outside their immediate area. Students are encouraged to find people in their families who come from diverse cities both within and outside the United States. As students bring in this information, you may want to record it on a U.S. or world map by having students place thumbtacks or pushpins on the map at the appropriate places. You can post beside the map a list of who comes from where. Students then work in groups to plan a trip to the birth cities of people in their families. Each team of students selects their cities, determines the distance between each pair of cities, and determines the distance for the entire trip.

■ Ask students to find the routes of famous journeys on U.S. or world maps and to find out the distances traveled on these journeys. You may have studied some journeys in social studies that students can explore. The lives of historical figures also present possibilities for tracing the paths of where they lived during their lifetimes. If you have studied the history of slavery or civil rights, students can look at some of the journeys of enslaved people attempting to get to freedom. A good source for students is *Many Thousand Gone: African Americans from Slavery to Freedom* by Virginia Hamilton (Knopf, 1993). One of the journeys in this book is that of Olaudah Equiana. He was born in 1745 in Benin, Nigeria; kidnapped at the age of 11; and sent on a slave ship to Barbados in the West Indies. He was sold several times, traveling from the West Indies to England, where he worked for two women who taught him to read; then he traveled to Philadelphia, where he eventually bought his freedom from a Quaker merchant. Finally, he returned to England, where he wrote a book about his life (published in 1789) and worked for the abolition of slavery.

■ There are also interesting journeys in nature. Every year, birds migrate many thousands of miles. *The Wheel on the School* by Meindert DeJong (Harper, 1954) is a story involving the storks' migration from South

America to Holland. The longest journey taken by a bird each year is the migration of the Arctic tern. Crossing back and forth between its breeding ground near the North Pole and its wintering grounds near the South Pole, this bird travels about 25,000 miles each year. Some students may be interested in finding out about the whooping crane. These magnificent birds are on the brink of extinction in the United States; only about 150 remain. One group of these birds migrates each year from their nesting grounds in Alberta, Canada, to the Aransas Wildlife Refuge near Rockport, Texas, where they spend the winter. A natural disaster during migration, such as a severe storm, could easily wipe out this flock. The migration of the monarch butterfly is another interesting story. Thousands of these butterflies move together in a large colorful mass, covering 1800 miles from northeast of the Rocky Mountains to central Mexico. You may have animals, birds, or insects in your area that migrate; find out where they go when they leave your area and have students trace their journeys on a map and calculate the distances traveled. A local wildlife group, such as the Audubon Society, can help you locate relevant information.

- Students can interview parents or relatives about long journeys they have taken, find their routes on a map, and calculate the distances of the journeys.

Likely or Unlikely?

Basic Activity

Students think about events in the world around them, considering which events are likely and which are unlikely to occur. They sort statements about events into the two categories: *Likely* and *Unlikely*. As they become familiar with these ideas, adding the categories *Very Likely* and *Very Unlikely* encourages students to make finer distinctions about the probability of these events. Students can also decide whether one event is *more* likely or *less* likely than another. We avoid the categories *Certain* or *Impossible* because students of this age can get into endless arguments about whether it's indeed certain that the sun will rise tomorrow, or whether it's genuinely impossible that a large white rabbit will serve lunch in the school cafeteria today!

Likely or Unlikely? involves students in considering the likelihood of the occurrence of a particular event. Ideas about probability are notoriously difficult for children and adults. In the early and middle elementary grades, we simply want students to examine familiar events in order to judge how likely or unlikely they are. Students' work focuses on:

- describing events with terms such as *likely*, *unlikely*, *more likely*, and *less likely*
- deciding what sorts of events in their lives are more and less likely

Materials

- Chart paper
- Strips of paper for students' statements
- Tape

The first time you do this activity, prepare some statements, naming events that are likely or unlikely, ahead of time. For subsequent sessions, students can write likely/unlikely statements as part of the Ten-Minute Math session, at home, or when they arrive in class. Or, you might ask a few students to prepare some ahead. To ensure a supply of both kinds of statements, ask each student to write two, describing one likely and one

unlikely event. Each statement should be written on a strip of paper that can be taped somewhere in a list. See Step 2 for sample statements.

Procedure

Step 1. On chart paper, start a list with two headings, *Likely* and *Unlikely*. If this is the first time you are doing the activity, discuss with students what these words mean and what kinds of things are likely and unlikely.

Step 2. Read, one at a time, statements of events that are likely or unlikely to occur. Following are some ideas to start with. The likelihood of some of these events is, of course, related to the characteristics of your community, the season, and so forth.

> One hundred cars will pass our school during the day today.
>
> An airplane will land on our school roof today.
>
> Half of the students in our school will stay home with colds tomorrow.
>
> A few students will stay home with colds tomorrow.
>
> The school cafeteria will be noisy today.
>
> Fewer than 20 people in [our town] will order a pizza today.
>
> It will snow here tomorrow.
>
> It will rain here sometime in the next two weeks.
>
> Scientists will discover that Earth is flat.
>
> Our class will get five new students before the end of the year.

Step 3. Students decide whether the event each statement describes is likely or unlikely. After some discussion, tape the statement strip under the appropriate heading. Because there will not be enough time to discuss a statement from everyone in the class, select a few and save the rest for the next Likely or Unlikely? session. You can keep the list posted in the classroom and add new statements each time you do the activity.

Continued on next page

Variations

Two More Categories: *Very Likely* **and** *Very Unlikely* After students have had some experience with the ideas of likely and unlikely, ask them to write some statements that are *very likely* or *very unlikely*. Discuss: "How is a statement that is *very unlikely* different from one that is just *unlikely*? How is a statement that is *very likely* different from one that is just *likely*? How many of the statements we have posted already fit into these new categories? Can you think of a way to change a *likely* statement into a *very likely* statement?"

Changing Likely to Unlikely Look at your list of likely and unlikely statements. Ask students to choose one statement and change it so that it would move to the opposite list. For example:

Unlikely: An airplane will land on our school roof tomorrow.

Change to likely: An airplane will not land on our school roof tomorrow.

Likely: The school cafeteria will be noisy today.

Change to unlikely: The school cafeteria will be quiet today.

Choose a few of these to discuss. Do other students agree that the statement that was at first likely is now unlikely?

More or Less Likely? Introduce the element of comparison with statements using *more likely* or *less likely*. For example:

It is more likely that it will rain tomorrow than that it will snow.

It is less likely that I will see a mouse on the way home than that I will see a dog.

As you or the students suggest such statements, discuss them. Do students agree with them?

Related Homework Options

Writing Likely/Unlikely Statements At home, students write statements to bring in for sorting during the next Ten-Minute Math session. You may want to provide a homework sheet with two sentences to be filled in.

It is likely that _____ .

It is unlikely that _____ .

After students have some experience with these ideas, you can add other sentences, making finer distinctions or comparisons:

It is very likely that _____ .

It is very unlikely that _____ .

It is more likely/less likely that _____

than that _____ .

Connections with Other Events in the Community Students might write statements of likely or unlikely events that occur in their community. For example, they might consider these statements:

The river will flood this year.

In a few years, we will have less polluted air in our city.

The trash in the park will be cleaned up by next Sunday.

The new school addition will be finished by September.

They may need to interview some people who know about these events to help them decide whether they are likely or unlikely. They may even be able to set into motion actions that could change the probability of some event, such as organizing a park clean-up!

The following activities will help ensure this unit is comprehensible to students who are acquiring English as a second language. The suggested approach is based on *The Natural Approach: Language Acquisition in the Classroom* by Stephen D. Krashen and Tracy D. Terrell (Alemany Press, 1983). The intent is for second-language learners to acquire new vocabulary in an active, meaningful context.

Note that *acquiring* a word is different from *learning* a word. Depending on their level of proficiency, students may be able to comprehend a word on hearing it during an investigation, without being able to say it. Other students may be able to use the word orally but not read or write it. The goal is to help students naturally acquire targeted vocabulary at their present level of proficiency.

We suggest using these activities just before the related investigations. The activities can also be led by English-proficient students.

Investigation 1

menu, restaurant, order

1. As you rub your stomach, express how hungry you are. Now tell the students that you are just about to enter a restaurant. Pantomime opening the restaurant door and looking around. Then seat yourself at a table with a menu. (If possible, try to use a menu that includes pictures of the various foods offered.)

2. Pick up the menu and begin to look it over. Contemplate aloud what you might order. ("Oh, there are so many choices on this menu. Hmm. Maybe I'll order the hamburger. No—maybe I'll order the chicken. No—I know—I'll order the spaghetti.")

3. Now show students the menu. Ask questions that challenge them to look at the different prices. ("What could you order from this menu that costs less than $10.00? What could you order from this menu that costs more than $8.00? What is the most expensive food a person can order at this restaurant?") Students can point to their answers if they cannot read or identify the actual items on the menu.

receipts, prices, clerk, grocery store, change

1. Show and identify six different grocery store receipts. Since grocery receipts are small and sometimes hard to read, either pass one out to each student pair or copy them onto the board or overhead. Point to where the total appears on each.

2. Give a student two dollar bills and four quarters and tell the rest of the group that this person is now the grocery store clerk. Pantomime taking groceries out of the cart as the "clerk" pantomimes ringing them up.

Investigation 2

mileage, miles, odometer

1. Draw a rebus sketch of a car on the board. Next to it, draw a quick sketch of the car's dashboard.

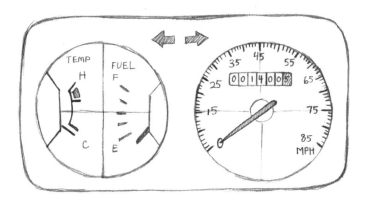

2. Explain that the part of the dashboard showing the car's mileage—or how much the car has been driven—is called an odometer. Ask students to read how many miles the odometer in the board sketch is showing.

3. Now sketch four more cars on the board. Tell the students that you are going to write the mileage under each sketch.

Blackline Masters

—————————————— , 19 ——

Dear Family,

In mathematics, we are beginning a unit called *Money, Miles, and Large Numbers* that has many connections to children's real lives. The unit is about adding and subtracting different kinds of numbers, with an emphasis on decimal numbers and large numbers. Children will start by working with money, then move on to work with distances. In order to understand how to make change, add up a restaurant bill, and calculate distances on a trip, children need plenty of experiences both in and outside of school. There are many ways you can be involved.

First, when you and your child are buying something, figure out together what the change will be. If you buy an item that costs $3.89 and give the clerk $5.00, figure out how much you should get back. When you buy several things, ask your child to help you estimate how much all the items will cost. Another place where children can practice adding and making change is at a restaurant. Ask your child to estimate how much the meal will cost. See if he or she can figure out the cost without using pencil and paper and ask the child to explain how he or she did it. Children will be doing this at school, and it would be helpful for them to have some real-life practice. Whenever these problems come up, encourage your child to figure out ways of estimating and making change that actually work. People do these problems in many different ways, and there's no one right way.

When we work on adding and subtracting distances, you can again help in several ways. Children need to get first-hand experience with distances like 1/10 of a mile, 1/2 of a mile, or a mile. If you drive, show them the odometer on your car and ask them to help you figure out how far it is to the grocery store or the playing field. This can be hard to do: "If you start at 24,532.1 miles, and when you get to the store the odometer reads 24,533.8, how far have you gone?" Encourage your child to find the answer mentally. On a trip, when kids ask "When will we get there?" you can answer that you will get there when the odometer reads a certain number. Then they can figure out how much farther it is. (This should keep them busy for a while.)

Another way children will get experience with decimals is by walking, running, or riding their bikes on routes where they know the distances. Help your child figure out how far it is to different places in your neighborhood. If your child walks to the store and back, how far has he or she gone? If there are any fun runs, bike rides, or walks for charities in your community, try to get involved in them. Many of these events involve distances with decimal amounts like 3.1 or 6.2 miles. These are excellent opportunities for children to get fit, to contribute to their community, and, of course, to become better mathematicians!

Sincerely,

Different Ways—Same Amount

Choose two grocery items. Record what they are and
what their prices are. Use play money to show the cost
in two different ways. Record both ways, drawing the
coins (and bills) below.

Switch items with your partner. On a new student
sheet, find two ways of showing prices for your
partner's items.

Compare your results. Have you and your partner
found the same ways or different ways? Have you or
your partner found the way that uses the fewest coins?

1. One item I chose is: _____

   Its price is: _____

   Two ways of showing the price are:

2. Another item I chose is: _____

   Its price is: _____

   Two ways of showing the price are:

What's for Lunch?

You have $6.00 for lunch. Choose one of the restaurant menus and order three or more things for your lunch.

Restaurant: _____

Lunch Order
(List the food and the price of each item. Then find the total cost.)

Total cost: _____

Change from $6.00: _____

First Grade Book Order

My Suggestions for a $100.00 Order

Page	Book Title	Price
Total		

What's in the Cupboard?

Choose one of the following amounts:

$3.00 $5.00 $10.00

Find grocery (or drugstore or hardware) items at home with prices that total *approximately* your chosen amount. List the items, their prices, and their total cost. Show how you found the total cost.

Close to 100 Score Sheet

Game 1   **Score**

Round 1: ____ ____ + ____ ____ = _____   _____

Round 2: ____ ____ + ____ ____ = _____   _____

Round 3: ____ ____ + ____ ____ = _____   _____

Round 4: ____ ____ + ____ ____ = _____   _____

Round 5: ____ ____ + ____ ____ = _____   _____

TOTAL SCORE _____

....

Game 2   **Score**

Round 1: ____ ____ + ____ ____ = _____   _____

Round 2: ____ ____ + ____ ____ = _____   _____

Round 3: ____ ____ + ____ ____ = _____   _____

Round 4: ____ ____ + ____ ____ = _____   _____

Round 5: ____ ____ + ____ ____ = _____   _____

TOTAL SCORE _____

Money on the Calculator

Figure out these problems in your head. Look carefully to see which numbers are dollars and which are cents. With a partner, read each problem aloud and say the answer. Then record how you would enter it on your calculator.

Add	Mental Answer	Enter on Calculator	Calculator Answer
30¢ and 70¢	_____	_____ + _____	= _____
$30 and 70¢	_____	_____ + _____	= _____
$3 and 7¢	_____	_____ + _____	= _____
$30 and 7¢	_____	_____ + _____	= _____
$9 and 8¢	_____	_____ + _____	= _____
$90 and 8¢	_____	_____ + _____	= _____
$90 and 80¢	_____	_____ + _____	= _____
$9 and 80¢	_____	_____ + _____	= _____
90¢ and 80¢	_____	_____ + _____	= _____
$90 and $80	_____	_____ + _____	= _____

Beat the Calculator (page 1 of 2)

Can you estimate each total (within $1.00) before your partner finds the total on a calculator?

1.

SPEEDY MARKET	
Oranges	$0.47
Raisins	$1.09
Cookies	$1.89
TOTAL	_____

2.

SPEEDY MARKET	
Apples	$1.47
Bread	$1.52
Milk	$2.59
Crackers	$0.88
TOTAL	_____

3.

SPEEDY MARKET	
Pineapple	$1.97
Cereal	$3.48
Butter	$2.39
Ice Cream	$2.65
TOTAL	_____

4.

SPEEDY MARKET	
Muffins	$1.32
Jelly	$1.97
Pears	$2.22
Soap	$0.56
Carrots	$0.25
TOTAL	_____

5.

SPEEDY MARKET	
Blueberries	$1.15
Soup	$0.32
Shrimp	$4.68
Cheese	$0.99
Soda	$0.30
TOTAL	_____

6.

SPEEDY MARKET	
Pudding	$0.26
Pudding	$0.26
Pudding	$0.26
Corn Chips	$0.72
Chicken	$3.49
Ketchup	$1.39
TOTAL	_____

Beat the Calculator (page 2 of 2)

Can you estimate each total (within $1.00) before your partner finds the total on a calculator?

7.

SPEEDY MARKET	
Pickles	$1.67
Mustard	$1.39
Ketchup	$1.09
Mayonnaise	$2.85
Lettuce	$0.55
TOTAL	_____

8.

SPEEDY MARKET	
Raisins	$1.49
Carrots	$0.59
Milk	$2.59
Yogurt	$0.69
Eggs	$1.15
TOTAL	_____

9.

SPEEDY MARKET	
Melon	$1.19
Chicken	$1.99
Pretzels	$2.09
Macaroni	$1.00
Cake Mix	$3.70
TOTAL	_____

10.

SPEEDY MARKET	
Bagels	$2.30
Sugar	$2.97
Butter	$0.82
Soap	$0.56
Tea	$3.25
TOTAL	_____

11.

SPEEDY MARKET	
Light Bulbs	$2.25
Bread	$1.52
Fish	$4.28
Applesauce	$1.29
Juice	$0.40
TOTAL	_____

12.

SPEEDY MARKET	
Soup	$0.32
Soup	$0.32
Soup	$0.32
Popcorn	$1.16
Rolls	$1.60
Sour Cream	$2.39
TOTAL	_____

One Lunch Order

Anna Maria ordered the following items for lunch:

a bean taco	$1.13
an order of rice	.89
a soda	.78

First make an estimate: About how much will Anna Maria's lunch cost? Explain how you made your estimate.

Exactly how much money will Anna Maria's lunch cost? How did you know?

If Anna Maria paid for her lunch with $5.00, how much change would she get back? Explain how you solved this problem.

How Much Change?

1.  $.59 $.29

How much change from:

$1.00? _____

$5.00? _____

$10.00? _____

2.  $.72 $.49

How much change from:

$1.50? _____

$5.00? _____

$10.00? _____

3.  $1.29 $1.17 $1.59

How much change from:

$5.00? _____

$10.00? _____

$20.00? _____

4. 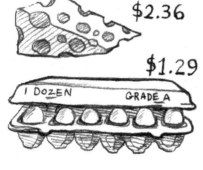 $2.36 $1.29    $.84

How much change from:

$5.00? _____

$10.00? _____

$20.00? _____

Camping Supplies

Annie went shopping and bought the following items to take on a camping trip:

bug spray	$1.79
batteries	$2.15
trail mix	$1.53
marshmallows	?
small compass	?
TOTAL	$7.68

1. Estimate how much Annie spent on the first three items. Write about how you got this estimate.

2. Annie spent $7.68 for her supplies. Figure out what the prices of the marshmallows and compass could be. Write a number sentence that shows the problem. Then write about how you decided on these prices.

3. Annie paid for her supplies with a $10.00 bill. How much change did she get back? Write a number sentence that shows this. Then write about how you solved that problem.

(Use the back of this sheet if you need to.)

Art Supplies

Annie went shopping and bought the following items for an art project.

paper	$2.79
glue	$1.29
glitter	$ .53
pencils	?
markers	?
TOTAL	$8.34

1. Estimate how much Annie spent on the first three items. Write about how you got this estimate.

2. Annie spent $8.34 for her supplies. Figure out what the prices of the pencils and markers could be. Write a number sentence that shows the problem. Then write about how you decided on these prices.

3. Annie paid for her supplies with a $10.00 bill. How much change did she get back? Write a number sentence that shows this. Then write about how you solved that problem.

CANTON EXPRESS

Soups

Egg Flower Soup	$1.23
Hot-and-Sour Soup	1.34
Won-Ton Soup	1.55

Entrees

Chicken Chow Mein	2.79
Moo Goo Gai Pan	3.59
Sweet-and-Sour BBQ Pork Ribs	3.79
Beef with Broccoli	2.98

Sides

Egg Roll	.95
Fried Won-Ton	1.19
Pork Fried Rice	1.45
Steamed Rice	.95

Beverages

Hot Tea/Coffee	.55
Iced Tea	.75
Soda	.85
Milk	.75

ANTOJITOS SABROSAS

Taco (chicken, beef, or bean)	$1.13
Burrito (chicken, beef, or bean)	1.59
Enchilada (chicken, beef, or cheese)	1.38
Tamale (beef or pork)	1.59
Quesadilla	1.13
Refried Beans	.89
Rice	.89
Tortillas de Harina (2 flour tortillas)	.69
Tortillas de Maiz (2 corn tortillas)	.49

Drinks

Lemonade	.78
Cola and Other Sodas	.78
Milk	.65
Fruit Juice	.93

SANDWICHES AND BURGERS

Sandwiches (White, wheat, or rye bread)
Peanut Butter and Jelly	$1.35
Egg Salad	1.45
Tuna Salad	1.55
Chicken Salad	1.65
Ham	1.75
Roast Beef	1.85
Corned Beef	1.95
Grilled Cheese	1.45

(Cheese can be added to the above for 15¢.)

Burgers
Basic Burger	1.35
Basic Burger with Lettuce and Tomato	1.45
Cheeseburger	1.55
Cheeseburger with Lettuce and Tomato	1.65

Side Orders
Potato Chips or Corn Chips	.58
Coleslaw	.78
French Fries	1.12
Apple, Orange, or Banana	.85

Drinks

Milk, Coffee, Tea, Lemonade, Cola, Root Beer, Lemon-Lime, Ginger Ale, Orange Soda

Sm. $0.45   Med. $0.65   Lg. $0.85

PIZZA AND PASTA

Pizza (by the slice)
Cheese	$1.28
Extra Toppings	.22 each

mushrooms, tomatoes, black olives, green peppers, onions, pineapple, pepperoni, sausage, salami, ground beef, anchovies

Pasta
Spaghetti with Tomato Sauce	$2.10
Spaghetti with Meat Sauce	2.56
Ravioli	2.37
Lasagna	2.88
Cannelloni	2.73

Extras
Green Salad	1.27
Garlic Bread	.88

Beverages
Soda	.80
Milk	.70
Iced Tea	.75
Coffee or Tea	.75

Investigation 1 • Resource
Money, Miles, and Large Numbers

Materials

- One deck of Numeral Cards
- Close to 100 Score Sheet for each player

Players: 1, 2, or 3

How to Play

1. Deal out six Numeral Cards to each player.

2. Use any four of your cards to make two numbers. For example, a 6 and a 5 could make either 56 or 65. Wild Cards can be used as any numeral. Try to make numbers that, when added, give you a total that is close to 100.

3. Write these two numbers and their total on the Close to 100 Score Sheet. For example: 42 + 56 = 98.

4. Find your score. Your score is the difference between your total and 100. For example, if your total is 98, your score is 2. If your total is 105, your score is 5.

5. Put the cards you used in a discard pile. Keep the two cards you didn't use for the next round.

6. For the next round, deal four new cards to each player. Make more numbers that come close to 100. When you run out of cards, mix up the discard pile and use them again.

7. Five rounds make one game. Total your scores for the five rounds. LOWEST score wins!

Scoring Variation

Write the score with plus and minus signs to show the direction of your total away from 100. For example: If your total is 98, your score is −2. If your total is 105, your score is +5. The total of these two scores would be +3. Your goal is to get a total score for five rounds that is close to 0.

Cereal	2.39
Tuna	.62
Bread	1.47
Milk	.75
Ice Cream	2.47
Cheese	.78

Cookies	1.99
Butter	1.39
Lettuce	.39
Corn	2.50
Oranges	2.45
Bananas	1.15

Investigation 1 • Resource
Money, Miles, and Large Numbers

COINS AND BILLS

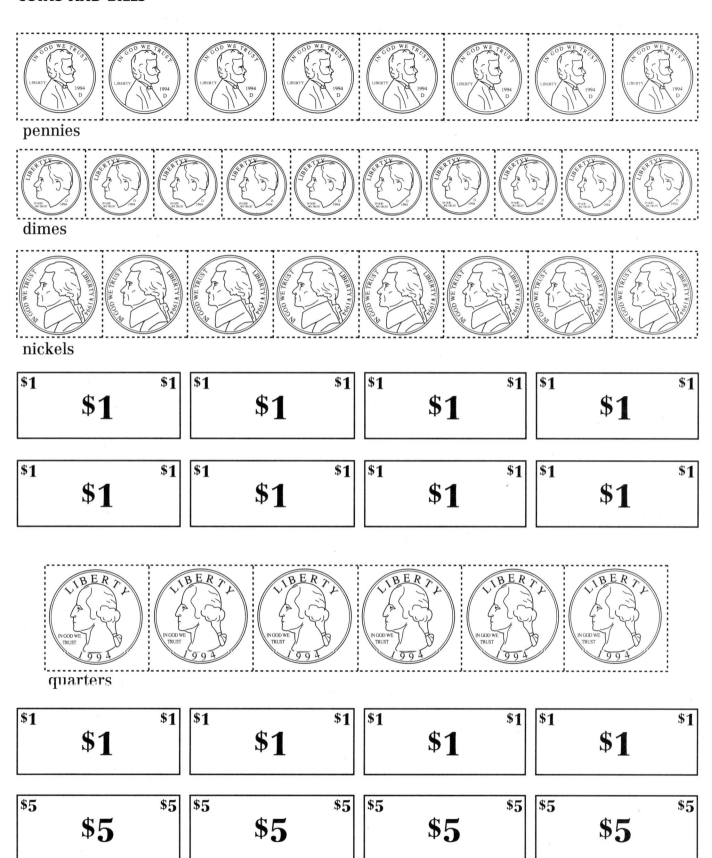

pennies

dimes

nickels

$1 $1 $1 $1

$1 $1 $1 $1

quarters

$1 $1 $1 $1

$5 $5 $5 $5

Runners' Logs (page 1 of 2)

For each running log, figure out if the person has gone
enough miles to prepare for a 3.1-mile race.
Remember: In a week, you should go at least 3 times
the distance you plan to race.

Sarah: 10-year-old; has run one shorter race before

Day	Mileage	Comments
Monday	2.2 miles	Ran around the pond once. I had stitches.
Tuesday	1.50 miles	I ran on the track, 6 times around.
Wednesday	1.25 miles	I ran on the track again.
Thursday	0 miles	I was tired and had to rest.
Friday	2.9 miles	Was visiting my aunt, ran with her.
Saturday	.8 miles	I was worn out from yesterday.
Sunday	1 mile	I ran it pretty slowly.

How much did Sarah run? Show how you figured it out.

Runners' Logs (page 2 of 2)

Nhat: $9\frac{1}{2}$-year-old; has run two races before

Day	Mileage	Comments
MONDAY	1.75 miles	I ran with my mom to the store, but we got a ride back.
TUESDAY	1.6 miles	Jamie and I ran to school because we were late!
WEDNESDAY	0 miles	I had to baby-sit today, so I couldn't run.
THURSDAY	3.2 miles	I ran slowly, but much farther than before.
FRIDAY	.5 miles	I was really tired, so I only ran around the track twice.
SATURDAY	1.75 miles	I ran home from the store, but slowly.
SUNDAY	.8 miles	Ran pretty fast, but not very far.

How much did Nhat run? Show how you figured it out.

Making a Running Log

Day	Mileage	Comments

Total miles for this week: _____

Calculator Skip Counting

Find some numbers less than 1 that you can skip count by to reach exactly 1, 2, or 3. Try it on the calculator. Write down the ways you counted.

Ways to Count to 1:

I counted by _____.   Here is the count:

I counted by _____.   Here is the count:

Ways to Count to 2:
(Can you find ways that skip 1?)

I counted by _____.   Here is the count:

I counted by _____.   Here is the count:

Ways to Count to 3:
(Can you find ways that skip 1 and 2?)

I counted by _____.   Here is the count:

I counted by _____.   Here is the count:

Investigation 2 • Sessions 1–2
Money, Miles, and Large Numbers

How Far Is $\frac{1}{10}$ of a Mile?

There are 528 feet in $\frac{1}{10}$ of a mile. To help you imagine how long $\frac{1}{10}$ of a mile is, you will measure it. Decide what you will measure with and how many of these it will take to make $\frac{1}{10}$ of a mile.

I used _____ as a measuring tool.

It was _____ long.

It would take about _____ of these to make $\frac{1}{10}$ of a mile.

It would take about _____ of these to make 1 mile.

Now decide where you will measure $\frac{1}{10}$ of a mile.

Make a map showing where and how you measured.

My Measuring Method

Use the following questions to write a brief report on the measuring method you used to measure $\frac{1}{10}$ of a mile.

What was your plan?

What did you do?

What did you find out?

On the back of this sheet, sketch a map showing your results.

A Tour of Our Town

These are the places we are including on our tour.
(Your tour should begin and end at school.) We will
visit them in this order:

1. _____   4. _____

2. _____   5. _____

3. _____   6. _____

From	To	Distance
1. _____	_____	_____
2. _____	_____	_____
3. _____	_____	_____
4. _____	_____	_____
5. _____	_____	_____
6. _____	_____	_____

The total mileage of our tour: _____

My Tour

List the places you will visit on your tour.

Write a description of the tour you have planned from the perspective of a tour guide. Describe each place the tour visits and what might be interesting or special about that place.

Close to 1000 Score Sheet

Game 1   Score

Round 1: ___ ___ ___ + ___ ___ ___ = _____ _____

Round 2: ___ ___ ___ + ___ ___ ___ = _____ _____

Round 3: ___ ___ ___ + ___ ___ ___ = _____ _____

Round 4: ___ ___ ___ + ___ ___ ___ = _____ _____

Round 5: ___ ___ ___ + ___ ___ ___ = _____ _____

TOTAL SCORE _____

....

Game 2   Score

Round 1: ___ ___ ___ + ___ ___ ___ = _____ _____

Round 2: ___ ___ ___ + ___ ___ ___ = _____ _____

Round 3: ___ ___ ___ + ___ ___ ___ = _____ _____

Round 4: ___ ___ ___ + ___ ___ ___ = _____ _____

Round 5: ___ ___ ___ + ___ ___ ___ = _____ _____

TOTAL SCORE _____

Travel Log: U.S. Trip

From	To	Length	Distance

How I figured out the distance between these two
places:

Comments:

Total Distance Traveled So Far:

Miles Left:

U.S. Map

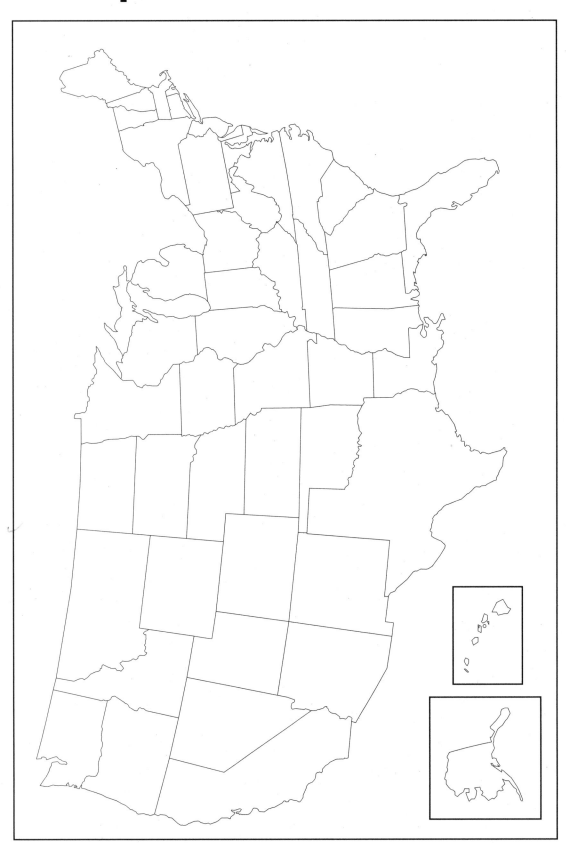

Materials

- One deck of Numeral Cards
- Close to 1000 Score Sheet for each player

Players: 1, 2, or 3

How to Play

1. Deal out eight Numeral Cards to each player.

2. Use any six cards to make two numbers. For example, a 6, a 5, and a 2 could make 652, 625, 526, 562, 256, or 265. Wild Cards can be used as any numeral. Try to make two numbers that, when added, give you a total that is close to 1000.

3. Write these numbers and their total on the Close to 1000 Score Sheet. For example: 652 + 347 = 999.

4. Find your score. Your score is the difference between your total and 1000.

5. Put the cards you used in a discard pile. Keep the two cards you didn't use for the next round.

6. For the next round, deal six new cards to each player. Make more numbers that come close to 1000. When you run out of cards, mix up the discard pile and use them again.

7. After five rounds, total your scores. Lowest score wins!

Scoring Variation

Write the score with plus and minus signs to show the direction of your total away from 1000. For example: If your total is 999, your score is –1. If your total is 1005, your score is +5. The total of these two scores would be +4. Your goal is to get a total score for five rounds that is close to 0.

0	0	1	1
0	0	1	1
2	2	3	3
2	2	3	3

4	4	5	5
4	4	5	5
<u>6</u>	<u>6</u>	7	7
<u>6</u>	<u>6</u>	7	7

8	8	<u>9</u>	<u>9</u>
8	8	<u>9</u>	<u>9</u>
WILD CARD	**WILD CARD**		
WILD CARD	**WILD CARD**		

Practice Pages

This optional section provides homework ideas for teachers who want or need to give more homework than is assigned to accompany the activities in this unit. The problems included here provide additional practice in learning about number relationships and in solving computation and number problems. For number units, you may want to use some of these if your students need more work in these areas or if you want to assign daily homework. For other units, you can use these problems so that students can continue to work on developing number and computation sense while they are focusing on other mathematical content in class. We recommend that you introduce activities in class before assigning related problems for homework.

101 to 200 Bingo   This game is introduced in the unit *Mathematical Thinking at Grade 4*. If your students are familiar with the game, you can simply send home the directions, game board, Tens Cards, and Numeral Cards so that students can play at home. If your students have not played the game before, introduce it in class and have students play once or twice before sending it home. You might have students do this activity four times for homework in this unit.

Ways to Count Money   This type of problem is introduced in the unit *Mathematical Thinking at Grade 4*. Here, three problem sheets are provided. You can also make up other problems in this format, using numbers that are appropriate for your students. Students find two ways to solve each problem. They record their solution strategies.

Story Problems   Story problems at various levels of difficulty are used throughout the *Investigations* curriculum. The three story problem sheets provided here help students review and maintain skills that have already been taught. You can also make up other problems in this format, using numbers and contexts that are appropriate for your students. Students solve the problems and then record their strategies.

How to Play 101 to 200 Bingo

Materials
- 101 to 200 Bingo Board
- One deck of Numeral Cards
- One deck of Tens Cards
- Colored pencil, crayon, or marker

Players: 2 or 3

How to Play

1. Each player takes a 1 from the Numeral Card deck and keeps this card throughout the game.

2. Shuffle the two decks of cards. Place each deck face down on the table.

3. Players use just one Bingo Board. You will take turns and work together to get a Bingo.

4. To determine a play, draw two Numeral Cards and one Tens Card. Arrange the 1 and the two other numerals to make a number between 100 and 199. Then add or subtract the number on your Tens Card. Circle the resulting number on the 101 to 200 Bingo Board.

5. Wild Cards in the Numeral Card deck can be used for any numeral from 0 through 9. Wild Cards in the Tens Card deck can be used as + or – any multiple of 10 from 10 through 70.

6. Some combinations cannot land on the 101 to 200 Bingo Board at all. Make up your own rules about what to do when this happens. (For example, a player could take another turn, or the Tens Card could be *either* added or subtracted in this instance.)

7. The goal is for the players together to circle five adjacent numbers in a row, in a column, or on a diagonal. Five circled numbers is a Bingo.

101	102	103	104	105	106	107	108	109	110
111	112	113	114	115	116	117	118	119	120
121	122	123	124	125	126	127	128	129	130
131	132	133	134	135	136	137	138	139	140
141	142	143	144	145	146	147	148	149	150
151	152	153	154	155	156	157	158	159	160
161	162	163	164	165	166	167	168	169	170
171	172	173	174	175	176	177	178	179	180
181	182	183	184	185	186	187	188	189	190
191	192	193	194	195	196	197	198	199	200

Practice Page
Money, Miles, and Large Numbers

0	0	1	1
0	0	1	1
2	2	3	3
2	2	3	3

Practice Page
Money, Miles, and Large Numbers

4	4	5	5
4	4	5	5
<u>6</u>	<u>6</u>	7	7
<u>6</u>	<u>6</u>	7	7

8	8	<u>9</u>	<u>9</u>
8	8	<u>9</u>	<u>9</u>
WILD CARD	**WILD CARD**		
WILD CARD	**WILD CARD**		

100

+10	**+10**	**+10**	**+10**
+20	**+20**	**+20**	**+20**
+30	**+30**	**+30**	**+40**
+40	**+50**	**+50**	**+60**
+70	**WILD CARD**	**WILD CARD**	**WILD CARD**

-10	**-10**	**-10**	**-10**
-20	**-20**	**-20**	**-20**
-30	**-30**	**-30**	**-40**
-40	**-50**	**-50**	**-60**
-70	**WILD CARD**	**WILD CARD**	**WILD CARD**

Practice Page A

Find the total amount of money in two different ways.

        3 pennies
        6 quarters
        3 nickels
        1 dime

Here is the first way I found the total amount of money:

Here is the second way I found the total amount of money:

Practice Page B

Find the total amount of money in two different ways.

> 8 nickels
> 9 dimes
> 2 pennies
> 7 quarters

Here is the first way I found the total amount of money:

Here is the second way I found the total amount of money:

Practice Page C

Find the total amount of money in two different ways.

> 1 half dollar
> 3 pennies
> 7 nickels
> 2 dimes
> 2 quarters

Here is the first way I found the total amount of money:

Here is the second way I found the total amount of money:

Practice Page D

For each problem, show how you found your solution.

1. Six friends go fishing. They go to a local bait shop to buy fish bait. They get 34 slugs. How many will each friend get if they want to share them equally?

2. Six friends together have 34 dimes, which they will share equally. How many dimes will each friend get?

3. Six friends decided to raise money by mowing lawns. They raised 34 dollars. They want to share the money equally. How much should each person get?

Practice Page E

For each problem, show how you found your solution.

1. I am planning a Hawaiian party. I will need 15 flowers for each lei that I make. How many flowers will I need for the 32 leis that I want to make?

2. There is a possibility that I will need 40 leis, instead of 32. How many more flowers will I need to make 40 leis?

3. How many flowers, altogether, will I need to make the 40 leis?

Practice Page F

For each problem, show how you found your solution.

1. The art teacher has 100 shells that he collected at the beach. His 24 students are making ceramic paperweights with shells. How many shells could each student put on his or her paperweights if the shells are shared equally?

2. The same art teacher has 200 beads. How many beads should he give to each of his 24 students for another project?

3. For yet another project, the art teacher has 300 stickers. How many should he give to each of his 24 students?